ENVIRONMENTAL CHANGE IN AUSTRALIA SINCE 1788

TO MY PARENTS,
Rua and Peter Johnson

ENVIRONMENTAL CHANGE IN AUSTRALIA SINCE 1788

Second Edition

ANN YOUNG

OXFORD
UNIVERSITY PRESS

253 Normanby Road, South Melbourne, Victoria 3205, Australia

Oxford University Press is a department of the University of Oxford.
It furthers the University's objective of excellence in research, scholarship, and education by publishing worldwide in

Oxford New York

Auckland Cape Town Dar es Salaam Hong Kong Karachi
Kuala Lumpur Madrid Melbourne Mexico City Nairobi
New Delhi Shanghai Taipei Toronto

With offices in

Argentina Austria Brazil Chile Czech Republic France Greece
Guatemala Hungary Italy Japan Poland Portugal Singapore
South Korea Switzerland Thailand Turkey Ukraine Vietnam

First published 1996
Reprinted with corrections 1998
Second edition published 2000
Reprinted 2002, 2004, 2005

National Library of Australia
Cataloguing-in-Publication data:

Young, Ann Rua Mackenzie, 1950– .
Environmental change in Australia since 1788.

2nd ed.
Bibliography.
Includes index.
ISBN 0 19 551292 8.

1. Human ecology—Australia.
2. Nature—Effect of human beings on—Australia.
3. Natural resources—Australia
I. Title.

304.20994

Typeset by Syarikat Seng Teik Sdn Bhd, Malaysia
Printed through Bookpac Production Services, Singapore

Contents

Figures, plates, tables and boxes

Figures

Colour plates

Tables

Boxes

Acknowledgments

I am very grateful to Jill Henry and the staff of Oxford University Press for their help and encouragement; to Robert Young for critically reading drafts of the revised manuscript; to Richard Miller for preparation of diagrams; and to Penny Williamson for photographic work.

I thank the following companies and organisations for permission to use copyright material:

Albatross Books for extract from Bruce Smith, 'Who knows', from his *I'll Not Pretend,* Albatross Books, Sydney, 1986

HarperCollins Publishers (Australia) Pty Ltd for extract from Judith Wright, 'Eroded Hills', from her *Collected Poems,* Angus & Robertson, Sydney, 1987

Plate C (vegetation maps 1780s and 1980s) courtesy Australian Surveying and Land Information Group, Canberra. © AUSLIG. All rights reserved.

Cooperative Research Centre for Satellite Systems for Plate D (clearing in the Nyngan area as indicated by satellite images) from Dean Graetz, Rohan Fisher and Murray Wilson, *Looking Back: The Changing Face of the Australian Continent, 1972–1992,* COSSA 029, CSIRO Office of Space Science and Application, Canberra, 1992. Landsat data produced by US Geological Survey and the Australian Centre for Remote Sensing, and processed by CSIRO.

Control Publications for Figure 3.3 (monetary analogy of soil erosion) from *Search: Science and Technology in Australia and New Zealand*

Foundation for a Rabbit-Free Australia for Figure 3.6 (anti-rabbit logo)

NSW Agriculture for Figure 4.3 ('a good burn', Dorrigo) from H. Jensen, *The Soils of New South Wales,* NSW Government Printer, Sydney, 1914

North Ltd, Melbourne, for Plate H (commercial forests, Tasmania)

The Editor, *Australian Geographer* for Figures 9.2 and 9.5 (wilderness quality in Tasmania) and 9.5 (Recreation Opportunity Spectrum)

WA Satellite Technology and Applications Consortium, and Satellite Remote Sensing Services, Department of Land Administration, Perth for Plate J (satellite image of Leeuwin Current)

Mitchell Library, State Library of NSW, for Figure 7.7 (flooded road, Maroubra)

Who knows

Who knows
where the river flows,
who knows?
Looking back
we can trace the bends,
but who knows
where the river tends,
let alone where the
river ends?
Who knows?

Bruce Smith 1986,
'Who knows' from *I'll Not Pretend,*
Albatross Books, Sutherland NSW

Preface to the second edition

Tracing the bends

THERE IS NO SHORTAGE of books proclaiming the imminent doom of the Australian environment, and denouncing the rape of the landscape that has occurred since White settlement. In the Foreword of a popular publication by William Lines (1991), David Suzuki thunders that the book 'puts the lie to the myth of the heroic history of modern Australia and reveals it as the sordid tragedy it really was'. Policy was driven by 'greed, shortsightedness and arrogance', in Suzuki's opinion. Contrary opinions are caricatured by including carefully chosen quotations, such as that of Lang Hancock, the mining magnate, who described environmentalists as 'subversives . . . threatening the lives and fortunes of the Australian community'.

This book covers the same period—the 200 years since 1788—but from a different perspective from that of Lines and Suzuki. I cannot feel the same sense of urgent and impending catastrophe that would appear to motivate Suzuki's diatribe. There is no doubt that environmental changes since 1788 have been dramatic and have affected the whole of Australia. There is no pristine and unmodified landscape left. However, the landscape is not totally despoiled and ravaged. A traveller through rural Australia will find that productive fields and pasture lands are more extensive than areas degraded by salinisation and gullies. The levels of pollution in Australian cities are rarely as severe as those in many cities in other parts of the world. This does not mean we can be complacent and assume that nothing needs to be corrected, but only that we need not overstate the severity of our problems.

However, perhaps the major difference in the perspectives of environmentalists such as Suzuki and myself can be illustrated by statements such as the following:

> Driven by a profound disconnection from the land, newcomers to the New World sought to tame it and its human and non-human occupants. The combined technology and the western attitude of rightful dominion over Nature, were unstoppable.
>
> Suzuki, 'Foreword', in Lines 1991

These assertions imply that people wilfully sought to destroy their natural environment and heedlessly modified their surroundings for entirely base motives: out of greed, shortsightedness and arrogance. They imply that motives such as these are unique to Western cultures, at least in relation to environmental modification, and that only indigenous peoples have any sense of affinity with the places in which they live. These are overstatements, yet we see them repeated and uncritically accepted, time and again. Even more scholarly works such as Aplin's (1998) introduction to environmental issues in Australia and

White's (1997) appraisal tend to be highly pessimistic and didactic. Associated with the ideas in these books are deep feelings of guilt for actions over which we have had no control, and of despair concerning problems for which we see little hope of solution. The best that seems possible is to slow the tide of degradation by educating the community to more responsible action.

We need, I believe, to avoid judging the actions of past generations by our present standards; to avoid replacing the 'three cheers' view of our history with the 'black armband' view (Blainey 1993). We also need to bring careful reason to any assessment of environmental damage, its extent and its causes. There is little point in attempting to manage better if we lack good understanding of the environment we are trying to control, yet rhetoric and emotive language are far more prominent than careful analysis in most environmental debates. The aim of this book is not to arouse fervour, but to present as objectively as possible the processes and changes which have shaped the Australian physical environment over the last 200 years.

I have confined the discussion to this period because the rate and extent of environmental change before and after White settlement differs so significantly. Firing of vegetation by Aboriginal people in the tropics seems to have caused regional changes in vegetation patterns soon after their arrival. Hunting may have led to extinction of megafauna, but in Victoria some now-extinct animals survived for more than 10 000 years after the arrival of humans. The relatively recent introduction of the dingo (about 6000 years ago) with a migration of newcomers led to the extinction of the *Thylacine* (Tasmanian tiger) on the mainland. In southern Australia and Tasmania, fire probably affected vulnerable communities and altered the vegetation mosaic at a local level, rather than having regional impacts on vegetation patterns. These changes (Dodson 1992) were significant, but occurred over tens of thousands of years, and are accepted by most researchers as being of a different dimension altogether from the more recent changes. The maverick with respect to these views is Tim Flannery, in his book and TV series 'The Future Eaters'.

Flannery (1997), in his introduction, rightly identifies a major change recently in our view of the continent, with an end to a British interpretation of land following native title recognition, a realisation of ecosystem damage, and a need to determine how many people the continent can sustain. Then he argues:

- that rainforest blanketed eastern Australia until the late Pleistocene 100 000 years ago
- that a number of cores attest to increased fire frequency/severity some time between 100 000 and 38 000 years ago
- that this is due to Aborigines causing the extinction of megafauna, and to the resulting vast increase in uneaten grasses providing huge fuel sources
- that consequent fires led to erosion, siltation downriver and mangrove expansion in north Queensland, and to the spread of heathland plants from areas of poor soils.

Plausible though this may sound, the edifice is built on meagre foundations. It rests on a small number of cores with rather sparse chronology, places contemporaneously a few pieces of evidence that many other researchers would not consider to be connected, and extrapolates well beyond the limits with which others would be comfortable. This raises a question that will be considered later: how do we assess the evidence we have for environmental impacts? Flannery's work suggests that Aboriginal occupancy changed the land far more than has been thought, but like others, he notes the quantum leap in impact since European invasion.

The drastic nature of recent changes is perhaps best seen in Australia's agricultural areas, and land degradation has properly become an issue of national political importance. Inappropriate policies and inadequate appreciation of the consequences of land-use changes led to a situation desperate enough to force together unlikely allies—the National Farmers Federation, the Australian Conservation Foundation, and mining companies such as Alcoa—to promote the Decade of Landcare. This is a positive move, and a move away from the trend against which N. F. Barr and J. W. Cary warn us in their account of the greening of our brown land:

> Our communal perspective of the land has been courted by advocates with two opposing systems of cultural values: environmentalism and economic rationalism. At one extreme of the environmental movement there is acceptance of a definition of land degradation that denies legitimacy to agricultural production itself ... From the [economic rationalist] viewpoint there is validity in the idea of an optimal rate of land degradation determined by society's preference for current and future income.
>
> We should not fall for the doomsday debating trick of reducing a continuum of possible policies to two extremes, and then showing the one extreme to be so unappealing that the alternative is seized on with grateful relief.
>
> Barr and Cary 1992, pp. 281, 286

As a summary of my approach, I cannot improve on the comments made by D. N. Jeans in his introduction to *Australia: A Geography*, which he edited first in 1977:

> Faced with the varied red, brown and green visage of Australia, its swarming cities and almost empty plains, its multiplicity of governments and uniform culture, geographers have a strong armoury of approaches and tools to bring to the task of analysis.
>
> [Geography recognises humankind as] a causal agent logically equivalent to natural forces ... It was this logic of the landscape view of geography that prompted the massive and cooperative *Man's Role in Changing the Face of the Earth* rather than any prescient knowledge of an environmental crisis, though this tradition leaves geography well-placed to engage in the current debate.
>
> Jeans 1977, pp. 4–5

Geographers are holistic by training and inclination, seeing both the human and the physical aspects of our world as interconnected. While my own expertise is more on the physical side, I have tried to take account of this integrated viewpoint. Like other professionals, we find it easier to interpret past events and causal links than to be truly prescient of the future. It is easier for us to recognise the mistakes of the past than to identify the best way forward for the future. Attitudes to the natural environment, and the socio-economic context of environmental management, have been constantly changing through history, and will continue to change. The events of the past give us some idea of what to expect in the future and how best to proceed, but we do not know exactly what lies ahead. Like a traveller on a river, we can trace the bends we have passed, but uncharted waters lie ahead.

Chapter 1

Changing Views of the Australian Environment

A harsh and unfamiliar land

IT WAS AN UNFAMILIAR LAND to which the British came in 1788. The seasons were topsy-turvy; the vegetation had evolved in a totally different floral realm, and largely in isolation, since the breakup of Gondwanaland; the trees were evergreen and not deciduous, and although many came from known botanical families, the genera were new. In the Sydney area, shrubs bore pear-shaped fruit that were solid wood; the native flowers were often inconspicuous and without perfume. One writer described the trees as a 'forest in rags' (Earle 1853, quoted in Jeans 1972, p. 55). There were no major rivers allowing easy access to the continent's interior, and the Blue Mountains formed an impenetrable barrier for 25 years. Yet the lack of major streams to aid exploration proved a hard fact to accept. Large rivers had allowed exploration and development in other new lands; surely it was only a matter of finding the Australian stream. As recently as 1830, it was suggested that a huge Mississippi-like river existed in Australia (Heathcote 1972). This river was thought to extend right across the continent, beginning west of the Great Divide and flowing into the sea south of the Kimberleys, having picked up another large tributary flowing up from southern Western Australia. The landscape was harsh and forbidding; the vegetation and soils were unproductive. As D. N. Jeans (1972, p. 80) comments, 'short rations and the constant threat of starvation daunted the settlers and left them feeling isolated and abandoned at the end of the earth'. H. Speirs (1981) quotes early explorers and travellers who described parts of the Blue Mountains in very derogatory terms. The town of Blackheath was an 'awful contrast to that beautiful place of that name in England', and a gully entered by Govett in 1835 was an 'infernal Hole'.

However, many of the early colonial paintings did not show a harsh landscape. The trees were neatly arranged and rounded, and the Blue Mountains resembled cleared hills (Figure 1.1). Even scenes painted between 1815 and 1820 showed fenced and orderly roads and pastures, occupied by well-dressed people, in parts of the Blue Mountains that had been crossed by a formed road only in 1815. This portrayed, but hardly realistic, order was an attempt to come to grips with an alien landscape:

Figure I.1 John Eyre's painting of Port Jackson Harbour, in Sydney, showing the Blue Mountains as cleared rolling hills, c. 1810.

> The artist's need for a visual order in the landscape no doubt stemmed from various sources including a somewhat desperate desire to make the best of the new 'home' of enforced exile by turning it into a blissful Arcadia . . . To impose an order was to find a sense of security through the return to images more reminiscent of Kew Gardens than of eucalypt forest with attendant understorey.
>
> Speirs 1981, p. 113

Australia was never 'Arcadia', an ideal pastoral countryside, where peace and prosperity reigned. Life in Australia was difficult and often tragic. The harsh conditions experienced by small settlers are evoked humorously by C. J. Dennis in his accounts of Dad and Dave's exploits at Snake Gully, and grimly by Henry Lawson in his short stories. While a preoccupation with 'the bush' might have originally been a clever promotional scheme for the *Bulletin* magazine, a publication that drew on the talents of writers such as Lawson and Banjo Paterson, the 1890s saw the cementing of the idea of the bush as valuable. Nevertheless, a sense of isolation and difficulty associated with living in the bush was experienced by explorers of the inland, as well as by small settlers, and has remained a persistent theme in the Australian consciousness well into the twentieth century. The novels of Patrick White, for instance, are powerful expressions of the theme (Brown 1991).

We could argue that the settlers came to grips with their new home, and imposed a new order upon it, only too quickly. They 'erupted west, north and south across the confining mountains' (Grenfell Price 1972, p. 91), seeking new pastures and establishing a way of life that excluded the Aborigines and caused dramatic modification of indigenous flora and fauna. Until the mid-1840s the rainforests of north-eastern Queensland provided a refuge for Aborigines. The forests were of no interest to the Europeans charting the coast, who described them simply as 'wooded hills'. After the ill-fated Kennedy expedition and the discovery of gold, the forests were occupied by Europeans. As they recognised its natural resources of timber and rich soils, they did not hesitate to clear what they perceived to be a hostile environment. It was hostile not only because of conflict with Aborigines, but because as 'jungle' it was seen as the source of 'miasmatic fevers' (Birtles 1997).

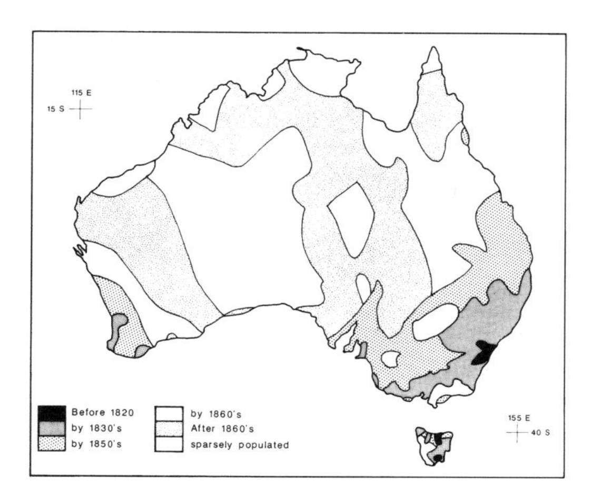

Figure 1.2 The spread of settlement in Australia, showing areas of early settlement in New South Wales and Tasmania, and the rapid spread of settlement across most of the well-watered land accessible from the southern coasts.

By 1850 settlement had spread into all of New South Wales, except the semi-arid western part (Figure 1.2).

The perception of the Australian landscape and its potential had changed substantially both in Australia and in Great Britain. The British no longer saw it as just as a penal colony. Now it was portrayed as 'a veritable Arcady in which the Golden Age of rural prosperity and individual dignity might be recaptured' (Powell 1977, p. 71). Promotion of the new settlements at the Swan River in Western Australia and at Adelaide under Wakefield's scheme in South Australia drew on this image. Unfortunately the survey of the Swan River by Captain Stirling in 1827 was done in a good season and did not extend into unfertile sand dune areas, so that comparisons made in England with Goshen, the Biblical land of plenty, and Hesperia, the Isles of the Blest in Greek mythology, turned out to be spurious (Powell 1977).

In Australia, people had come to appreciate the landscape in its own right. Colonial painters such as Conrad Martens and Eugene von Guerard were celebrating the Australian wilderness, choosing 'not to record the destruction of the wilderness, in part because of the unattractive character of such landscapes, and also because the romantic sentiment of their paintings would have suffered' (Bonyhady 1985, p. 79). Paintings showing the Zig-zag Railway near Lithgow on the western side of the Blue Mountains celebrated the crossing of this major physical barrier, but they differed from the earlier paintings of the road crossing in a number of significant ways. The structure itself was seen as striking and interesting, and the bareness of the construction area was clearly depicted. This is evident in Conrad Martens's paintings from the mid-1870s. Later tunnelling work at Lapstone is depicted by Arthur Streeton in his famous 'Fire's on!' painting, done in 1891. In this the excavated rock is an important feature of the scene, and the body being carried out from the tunnel is a symbol of the dangers of the battle between people and their new landscape. Gone are the unrealistically gentle and cleared pastures depicted by earlier painters. The painters of the late nineteenth century, from the 1870s onwards, aimed for scientific realism. They enjoyed friendships with leading geologists and botanists: Martens with W. B. Clarke in New South Wales; Chevalier with Julius Haast, and von Guerard with von Hochstetter in New Zealand; Piguenit with

James Scott and Robert Johnson in Tasmania. Von Guerard particularly was praised for the accuracy of geological and botanical detail in his paintings (Bonyhady 1985). R. L. Heathcote (1972) notes the strong and early interest in scientific description in the colonies. He suggests that this interest has persisted and remains a powerful argument for nature conservation measures.

Expansion and its consequences

The beginning of gold fever in 1851 drew numerous immigrants, including the first large contingent of non-British settlers: the Chinese. Population increased rapidly and nearly doubled in New South Wales between 1851 and 1861 (Jeans 1972). Demand for food increased, and after a brief period of labour-shortage problems, both pastoralists (especially those with beef cattle, rather than sheep) and cultivators found that the larger market allowed great expansion. Not all of this expansion was wise. In South Australia the demand for food from the Victorian gold-fields prompted the expansion of agriculture north towards Burra. A line was drawn by Goyder, the Surveyor General of South Australia, in 1865 to mark the 'safe' northern limit for agriculture. In 1872 all land up to Goyder's Line was opened up to farmers, and settlement surged northwards in the following good seasons. It soon passed Goyder's Line, with settlers proclaiming that 'the rain follows the plough'. The 1880–81 drought quickly destroyed this optimism, and by 1900 many farmers had abandoned their farms in the North (Williams 1969). But the experience was soon forgotten. M. Williams records the expansion past Goyder's Line into the Eyre Peninsula and the mallee country along the Murray River after 1900, and the subsequent crop failures and economic hardship after droughts in 1914–17. Only government assistance prevented widespread abandonment of these areas. It was not until 1935 that the State's Director of Agriculture advocated the return of land beyond Goyder's Line to sheep and cattle. He admitted that the Line really was the utmost limit of practical cultivation; it should never have been passed by wheat farmers (Williams 1974) (Figure 1.3). Bankruptcies and land degra-

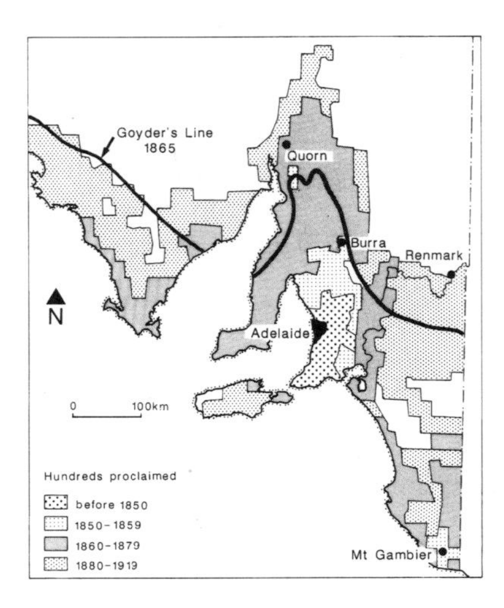

Figure 1.3
The spread of settlement in South Australia, showing expansion beyond Goyder's Line. Hundreds were a basic unit of land subdivision and were 100 square miles (approximately 270 km^2) in area. (Source: redrawn from Williams 1969.)

dation were the consequences of the unwise expansion beyond the boundaries of physically suitable land. Preoccupation with settlement of rural areas in both South Australia and New South Wales persisted after the First World War, demonstrating that 'political pressures for rural settlement were more powerful than were poignant but receding memories' (Wood 1997, p. 463).

In the gold-fields themselves the environmental consequences of development were severe. Lands were rapidly cleared of trees to provide dwellings, pit props, support structures and fuel. No thought was given to the effects of erosion and soil disturbance, and fertile floodplains were washed downstream to expose gold-bearing gravels below them. The area shown in Figure 1.4 was described by the Inspector of Mines as 'worthless swamp before it was mined'; it is certainly worthless swamp since it was mined. Mullock heaps of debris dragged out of shafts, and sterile mounds from the tables of alluvial workings, were abandoned and persist to this day near Tumbarumba, NSW, and in many other old gold-working areas (Plate A, Figure 8.1). Sluicing operations were also used in tin mining both in Tasmania and on the Atherton Tableland in Queensland, with similar dramatic impacts (Figure 1.5).

Figure 1.4
A remnant column of sediment and weathered granite, indicating the depth of material lost from the floodplain as a result of alluvial gold mining near Tumbarumba.

We should not suppose, however, that all people were blind to the consequences of settlement. Many settlers and government officials expressed concern about forest clearance, and the first timber reserves were set aside in New South Wales in 1871. This concern was due partly to the wide acceptance of the conservation ideals of G. P. Marsh, the influential American author of *Man and Nature* (Powell 1976). But there was no detailed or coordinated program of conservation, merely individual actions and isolated control measures by governments. J. G. Mosley (1972) comments that one of the most important of these measures in the long term has received almost no attention: in the 1860s a strip 100 feet wide was reserved along the coast for public use, and the beach culture of Australia today would be very different if that had not been done.

Figure 1.5
The sluice line at Glutton Gully tin mine, near Herberton, Queensland, c. 1910

The latter half of the nineteenth century saw the setting up of a large number of public reserves for recreation and scenery protection. Australia's first national park—Royal National Park, south of Sydney—was proclaimed in 1879. This occurred only seven years after the proclamation of the first national park in the world: Yellowstone in Wyoming. South Australia's first national park, the Belair Recreation Park, followed soon after in 1891. Like Royal National Park, it was initially seen more as a recreation area than as an area solely for nature preservation. There was still strong support for the introduction of European plants and animals, and Acclimatization Societies were set up to introduce desired species: larks and blackbirds for their song, trout for sport fishing, privet and hawthorn for hedges. Yet not all the efforts of these societies were detrimental. As Hutton and Connors (1999) note, the societies aimed for wise resource use and sustainable farming, and played an important role in promoting public awareness as well as exerting political pressure. Nevertheless, local critics argued that introduced species usually became pests, and there was growing appreciation of the native biota. By the 1890s all the mainland colonies had active Field Naturalist Clubs, and in South Australia the club's efforts led to the passing of the *Flora and Fauna Reserves Act* in 1919, establishing reserves in the rugged Flinders Chase area on Kangaroo Island, and at several places in the Adelaide hills (Whitelock 1985). The campaign to have a fauna and flora reserve on Kangaroo Island was a direct response to the playground function of Belair Park over-riding any nature conservation function (Hutton and Connors 1999). Today Flinders Chase remains an area of largely unspoiled vegetation, and much of the ecotourism pressure on Kangaroo Island is focused on the sea lion, seal and bird colonies on the southern coast (Plate B).

The establishment of Royal Societies also indicated a growing interest in native flora and fauna (Mosley 1972). Tasmania was the first to have its own Royal Society in 1842, followed by Victoria in 1859, New South Wales in 1866 and South Australia in 1876; most of these succeeded older Philosophical Societies; the Linnean Society of New South Wales was incorporated in 1884. There was a strongly pragmatic basis for much of this interest.

Goyder had marked out the limit of land suitable for agricultural settlement north of Adelaide on the basis of changes in the natural vegetation, reasoning that these reflected the overall climatic environment. Many other scientists recognised that the natural vegetation indicated the quality of soil and/or climate. Among them was Baron Ferdinand von Mueller, the Government Botanist in Melbourne from 1851 to 1873, who recognised that forest clearing would lead to sterility of soils, emphasised the usefulness of indigenous plants, and pioneered the use of eucalypts in revegetating arid areas in many parts of the world (Kynaston 1981).

The period also saw major expansion of settlement based on primary industries, and the revolutionising of transport, both on sea and on land, as a result of the steam engine (Jeans 1972). In New South Wales by 1900 the railways penetrated north to the Queensland border and Moree, north-west to Bourke, south to Cooma, and south-west to Albury and along the Murrumbidgee. The southern and south-western lines competed with Victoria for the Murrumbidgee wool clip and with the Murray riverboat trade, and followed a rapid expansion of wheat growing on the central and south-western slopes (Jeans 1972). Agricultural development dramatically and rapidly altered the pre-European landscape.

Hence in 1892 Hamilton could report to the Royal Society of New South Wales that one third of the forest in New South Wales had been cleared since settlement, that extension of *Callitris* pine scrub was a problem in western areas, that the number of rabbits in the colonies was 'astounding', that overstocking had destroyed pasture and that soils had lost nutrients 'exported' in meat, fleeces and hides. Selective use had reduced the numbers of many species: turpentines used for jetties, red cedar used for furniture and building, and wattles, whose bark was used in tanning, were no longer abundant in the Illawarra. He listed 165 species of introduced plants that were well established in Australia: 138 of these were found in New South Wales, 95 in Victoria, 79 in South Australia, 73 in Queensland, 64 in Tasmania and 13 in Western Australia. Not all were weeds. About 25% belonged to the *Gramineae* and *Leguminosae* families, and many of these were introduced as fodder plants. Among the weeds, blackberries and Scotch thistles gained special mention.

Others echoed Hamilton's comments. Holze (1892) listed over 60 species of plants introduced into the Darwin area, noting that they were occasionally useful fodder but generally were 'exterminating the native vegetation'. Also in 1892, a pastoralist of 30 years' experience, Samuel Dixon, delivered the following broadside:

> Necessarily as cultivation proceeds a larger area becomes despoiled of all the native plants, the forests are cleared, and repeated ploughings complete the extermination . . . The farmer, the squatter, the miner and the swagman all cause extensive conflagrations . . . [in the Riverina and western and northern South Australia, overstocking, drought and rabbits combine so that there is] for miles back from the river frontages, and in the neighbourhood of wells and dams, an unproductive surface, trodden down until almost impervious to water.
>
> Dixon 1892, pp. 196, 197, 202

Only in areas unsuited to agriculture, such as the 'unique and singularly lovely' sand plains of Western Australia, could Dixon imagine that the natural vegetation would survive. Hamilton, however, had one suggestion for the future that, he believed, was positive. Despite his concern about clearing and the introduction of exotic plants, he believed the introduction of irrigation would be of great benefit to the country.

Only water is needed?

Not everyone agreed with Hamilton's view. In the 1880s two Victorian investigators cautioned that Australia was not like other places where irrigation was successful, as its population was small and stream flow in summer was discontinuous. This and similar warnings were ignored in both Victoria and New South Wales. In Victoria the enthusiasm of Alfred Deakin for irrigation was decisive, and he introduced legislation to establish irrigation trusts. A severe drought in 1902 prompted interstate and Commonwealth investigations into the possibilities of irrigation along the Murray, and the establishment of the River Murray Commission in 1915 began the regulation of that river. New South Wales had already begun schemes for irrigation along the Murrumbidgee with the construction of Burrinjuck Dam. Enthusiasm for irrigation has continued largely unchecked, despite early and continuing evidence that the schemes are unprofitable and the development of salinisation is inevitable. The case against irrigation is argued most persuasively by Bruce Davidson (1969), who comments:

> No belief is more firmly held in Australia than that the scarcity of water has seriously hindered the nation's development . . . In spite of this belief it is simple to demonstrate that none of Australia's irrigation schemes operates profitably and that the Australian people would have had a higher standard of living if the area of irrigated land had been smaller.
>
> Davidson 1969, pp. 1, v

These are strong words indeed. And the irrigation issue illustrates well the power of community perceptions to direct public policy and consequently affect the Australian environment. Between the Second World War and the mid-1960s the area of land irrigated in Australia increased threefold (Davidson 1969), due to major government schemes that went ahead despite strong evidence of the risks of salinity problems and poor returns on the investment (Barr and Cary 1992).

The tension between recognition of physical restraints to growth and the assumption that only water was needed to allow almost unlimited growth has been a persistent theme in Australian environmental history. In the 1920s many people had unrealistic ideas about the value of much of the continent, and there was bitter public debate about the issue (Powell 1988). Griffith Taylor, a geographer with the newly formed Commonwealth Weather Service, suggested that Australia would support a maximum of 20 million people by the end of the century, and that all suitable land had already been settled. He rejected the idea of northern development as a 'white elephant' and described much of Western Australia as 'almost useless'. This was not a popular idea in a State where secessionist claims were being based on comparisons of the size of the State with the size of European countries (Figure I.6). Others saw Australia as a place of unlimited opportunities and suggested future Australian populations of over 60 million. These ideas lost favour by the end of the 1930s and Taylor's more realistic assessments were widely accepted. Nevertheless, as Powell comments in his detailed account of the controversy, 'Australia's leading politicians were not prepared to face up to environmental realities, and few of them had the capacity for seeing the continent as a whole' (Powell 1988, p. 146). Perhaps that is why A. Grenfell Price felt the need to write, some 40 years after Griffith Taylor's controversial statements:

> One point cannot be overstressed: the territorial area [of Australia] gives little indication of the habitable potentialities or value . . . If we subtract . . . useless Antarctic desert and . . . arid and semi-arid Australian country which cannot in general support dense populations, there remains, if one includes Papua-New Guinea, . . . an area considerably less than the size of the Argentine.
>
> Grenfell Price 1972, p. xiii

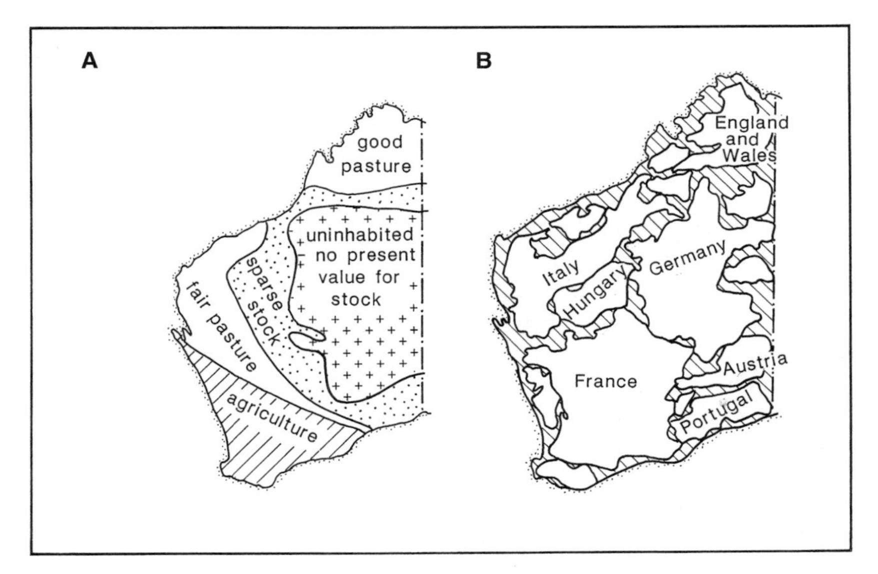

Figure I.6
Two contrasting views of Western Australia's productive potential:
A Griffith Taylor's assessment, showing large areas of 'useless land'
B The Western Australia Secessionists' Society's view, implying that Western Australia could be as productive as western Europe.
(Source: redrawn from Powell 1988.)

The optimistic view that only water is needed to make the Australian inland flourish is deeply ingrained in Australia's national culture. This is accentuated by the fact that Australia is one of the few Western democracies to have sparselands large enough to influence national policy. Of those few, it is the only country in which aridity, rather than cold climate, is the major restriction on development. This environmental constraint leads to locational constraints; although the wet–dry tropics of northern Australia have fertile soils, their remoteness meant that development schemes such as Humpty Doo and the Ord would have had to struggle for success, even if pests (magpie geese in the rice fields at Humpty Doo, and boll weevil in the Ord cotton crops) had not assailed them (Holmes 1988). Davidson (1969) maintains that the Ord scheme was unprofitable from the start and should not have been taken to completion. Indeed he contends that even the jewel in Australia's post-war engineering crown, the Snowy Mountains scheme, detracted from the nation's wealth rather than enhancing it.

Clearly, not all experts accept Davidson's opinions. C. H. Munro (1974) offers a rebuttal of the economic argument against irrigation and believes that water shortage and unpredictability is a challenge to be overcome, rather than a constraint about which Australians should become despairing. He points to the success of the Dutch in land-reclamation and of the Israelis in making the desert bloom as examples of how scientific and technological knowledge can conquer difficulties. He does comment disapprovingly on one plan to divert water inland—the Bradfield Scheme to divert flow from the Burdekin, Herbert and Tully Rivers on the Queensland coast into the Thompson and Flinders catchments—because it was based on 'inadequate topographic data' and this means that some aspects were seen to be impracticable. Almost in passing, he comments on the dam that was to change environmental attitudes in Australia more profoundly than any other development:

> Conservationists are now protesting against the inundation of Lake Pedder, claiming that this involves the destruction of an environment that has been very popular with nature lovers. It is an intangible cost. They claim that the objective of preserving Lake Pedder in its natural state is a more important objective than that of obtaining more hydro-power. Such controversies may occur frequently in the future.
>
> Munro 1974, p. 164

Water shortages and the effects of changing vegetation patterns have not been the only problem: the introduction of farming methods from Great Britain caused havoc in the vastly different Australian soils. Barr and Cary (1992), in their comprehensive account of this issue, describe the impact of early 'anarchic pastoral expansion' as 'devastating'. Subsequent soil depletion under cropping and pastoralism, as well as expansion into marginal lands, led to 'the erosion decades of the 1930s and 1940s'. There was some legislative response to the problems. South Australia passed its *Sand Drift Act* in 1923, and New South Wales enacted the *Soil Conservation Act* in 1938. As Mosley (1972) comments, this recognition was a late development in a country where the problem had become apparent shortly after settlement. Erosion control, new legumes, and rotational cropping led to a general increase in agricultural yields and a spirit of optimism after the Second World War. The trend towards pasture improvement gathered momentum in the 1950s, so that grazing land was cropped, hilly land cleared, and trees removed extensively. The optimism was ill-founded in areas that were only marginally suitable for these activities. Barr and Cary (1992) comment that many of Western Australia's current salinity problems are due to this relatively recent clearing of marginal lands. Certainly the recognition of problems such as salinisation and erosion has quelled the optimistic and expansionist views of the 1960s.

The move towards conservation

By the 1970s a mood of pessimism had overtaken many Australians, and there was an appreciation of the finitude of resources. This reflected a trend overseas, where the writings of the Club of Rome in its publication *Blueprint for Survival* and the writings of Paul Erhlich had aroused concern about rapid depletion of the earth's resources. E. Woolmington (1972) discusses the environmental constraints in terms of their social consequences. Aridity and distance had led to a high degree of urbanisation clustered on the coastal fringe. Rural communities have become frustrated with the drift of population to these cities, and governments have responded with policies of urban decentralisation; but this in turn has generated environmental problems, such as pressure on the water resources of the Murray River as a result of the Albury–Wodonga development.

However, it was events in Tasmania that finally turned the tide of public opinion about the consequences of developing Australia's natural resources. What J. Kirkpatrick (1988) calls the 'greening of Tasmanian consciousness' was also the greening of the mainland consciousness. The flooding of Lake Pedder in 1973 was not prevented by conservationists' protests. But after the Lower Gordon River Scheme was proposed in 1979 opposition to flooding of wilderness in order to provide hydroelectricity reached a crescendo. The Commonwealth government, with a clear mandate from the electorate—the 1983 election was fought to a significant extent on this issue—used the *World Heritage Properties Conservation Act 1983* to prevent the dam being built (Porter 1985). The 'No Dams' campaign that had begun with Lake Pedder meant that water storage no longer automatically received public approval in Australia.

The debate was, as Munro noted in 1974, not just about science, but about intangibles. Indeed, D. Mercer (1995) argues that values, and not facts, drive the central environmental issues of our time and that science does not have a good record in predicting significant environmental problems. There was an almost religious fervour within parts of the Australian community about the Tasmanian issue. The word 'wilderness' came to mean something that was desirable and should be preserved, instead of something threatening and unproductive that should be conquered. While scientific evidence against the hydroelectric scheme was gathered, the idea that science, or even economic advantage, could on their own justify development was rejected by many people; conservation values were seen as more important. It marked the beginning of the Green political movement in Australia. The United Tasmania Party, formed to oppose the Lake Pedder scheme, was, in fact, the first 'Green' party, and the failure to stop that scheme brought about important political changes: the Commonwealth government passed its 1974 *Environmental Protection (Impact of Proposals) Act*; the leadership of the Australian Conservation Foundation changed, and the Foundation began more direct political activity; conservation groups such as the Australian Conservation Foundation and the Wilderness Society campaigned on behalf of the Labor Party in the 1983 federal election. After 1983 the environment was placed firmly on the political agenda in both State and federal systems.

The Gordon-below-Franklin controversy also brought the environmental debate as a whole into the federal arena, although natural resource development is primarily a State right under the 1901 Constitution. Use of the World Heritage legislation has given the Commonwealth significant control of about 25% of Tasmania, and of extensive areas in other States also: for example, Kakadu in the Northern Territory, the Tropical Forests and Great Barrier Reef in Queensland, and the East Coast Rainforest Parks in New South Wales (Figure 1.7). States have followed the Commonwealth's lead in issues of global significance. State and Territory legislation is enacted to mirror Commonwealth legislation on matters such as reduction of ozone-depleting CFCs. Intergovernmental committees at ministerial level (such as the National Environment Protection Council) negotiate Australia-wide standards or approaches to environmental matters. Cooperation on matters such as national pollution standards would have been inconceivable ten years ago. While it would be wrong to think that these moves meet with unqualified approval throughout Australian society, or that all groups feel they are adequate, they are evidence of the strong political significance of environmental matters, and much of this significance can be traced back to the controversies in south-west Tasmania.

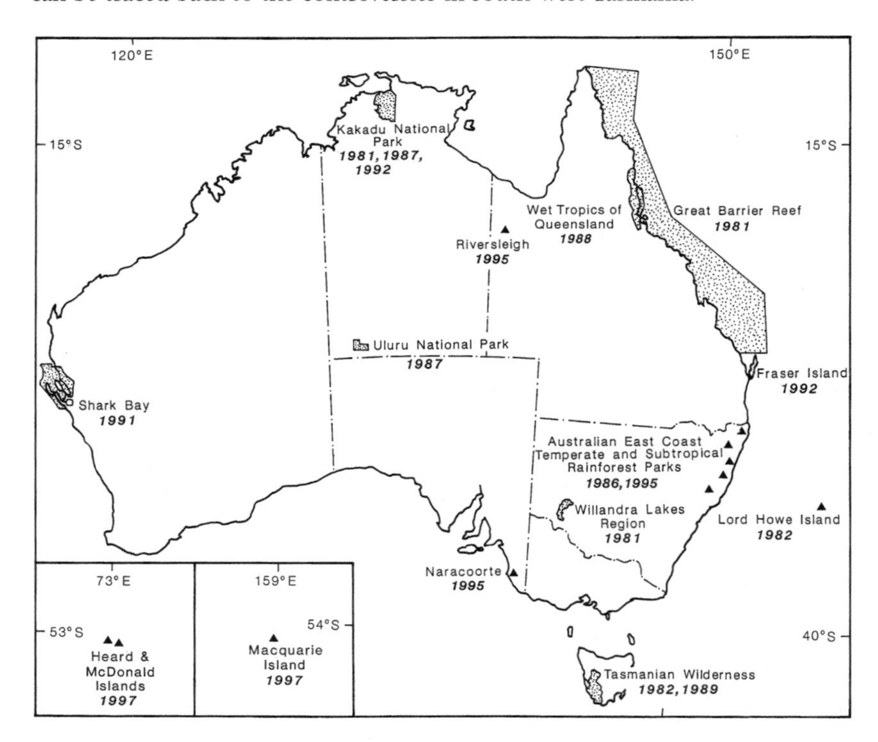

Figure 1.7 World Heritage areas in Australia and their dates of declaration.

Of course, these controversies were not the first 'conservation' issues of national significance. The cessation of whaling in Australian waters in 1980 and the prevention of mineral sand mining on Fraser Island in 1976, both brought about by the Commonwealth Government, were triggered by national outrage. In the case of Fraser Island, the Government had to use powers under its responsibility for customs, and therefore export licences, to achieve its goal (Environmental Defenders Office NSW 1992). The demand for the Fox Commission of Inquiry into the Ranger uranium project in 1975 was probably more the result of anti-nuclear feeling than of direct concern about what we may now term environmental considerations. Hence the findings of the Commission—which was constituted under the Commonwealth's *Environmental Protection (Impact of Proposals) Act 1974*—were reported in two volumes, one dealing with the issue of nuclear power worldwide and the other with local impacts on pastoralism, conservation and tourism (Porter 1985). Nor were the Commonwealth Acts the first significant environmental legislation to be enacted in Australia. By the early 1970s all States had passed legislation to require control of air and water pollution, and by the early 1980s environmental impact assessment procedures were established. However, the politicisation of environmental issues since 1983 clearly reflects a significant change in attitudes towards the environment among the Australian people. These ideas will be explored further in Chapter 9; but we will now turn to the question of how to understand and measure the changes that have occurred in the Australian landscape since 1788.

Chapter 2

Assessing Changes in the Australian Environment since 1788

Introduction

AS DISCUSSED in Chapter 1, it was an unfamiliar and topsy-turvy land to which the British settlers came in 1788. Incorrectly, as we know now, it seemed to them to be vacant land, a land uninhabited and of no account, a *terra nullius*. Into this they could move and establish a new branch of their civilisation, with minimal regard for the Aboriginal people whom they saw as scattered wanderers. The physical characteristic of the land that struck them most forcibly was its barrenness and aridity. Charles Oxley in 1817 described the semi-arid plains west of the Great Divide, around the Macquarie and Lachlan Rivers, as wearisome, wretched, barren, and uninhabitable by civilised man (Brown 1991). Yet not long after their discovery these same plains became the centre of the pastoral industry in New South Wales. Was the aridity real or imagined?

Perception and reality: Aridity in Australia

There is no doubt that Australia is an arid land. Of the inhabited continents, it has the lowest average annual rainfall and the lowest percentage of rainfall that runs off as surface water (Table 2.1).

Table 2.1
Rainfall and runoff in Australia and other continents

Continent	*Mean annual rainfall (mm)*	*% mean annual rainfall as runoff*
Australia	465	12
Antarctica/Greenland	150–200?	uncertain
North America/Europe/Asia/Africa	600–690	38–52
South America	1630	57

Data from Pigram 1988, Table 12.1, p. 152

Australia has dry conditions and clear air for much of the time because it lies under the subtropical subsidence zone and because it has low altitude over most of its area. Even the monsoon that brings rain to the tropical regions is shallower and less certain than its Northern hemisphere counterpart, and its penetration inland is variable. Potential evaporation is high (500–3000 mm annual average), and at Alice Springs, the total average annual rainfall could be evaporated by the radiation received in January. We can see this more clearly if Australia is drawn upside down, latitude for latitude, in the Northern hemisphere across Africa and Europe. Then much of the continent sprawls across the Sahara, the Middle East or the Gobi desert (Figure 2.1).

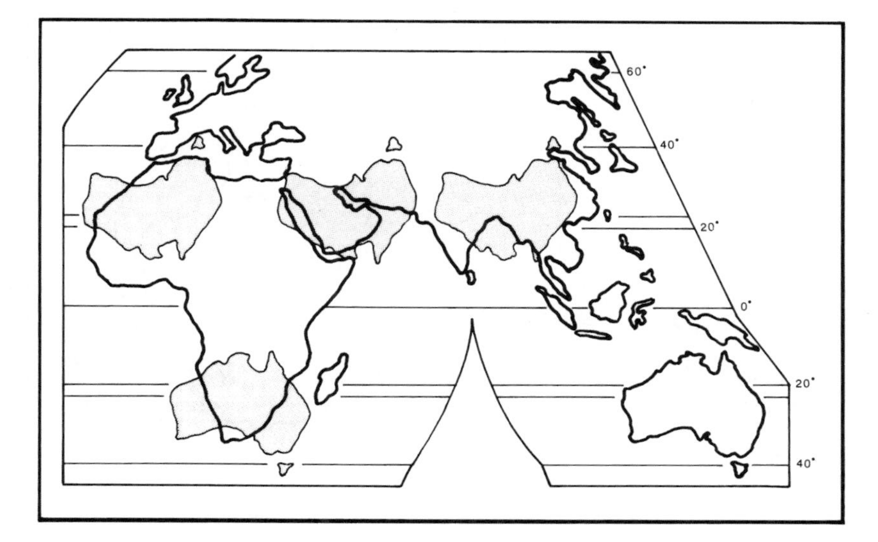

Figure 2.1
The relative latitudinal position of Australia, showing how this corresponds to the location of major desert areas in other parts of the world. (Source: redrawn from the *Macquarie Illustrated World Atlas* 1984.)

Nevertheless, some people consider scarcity of water in Australia to be an illusion (Warner 1986), and relative to population, Australia does have large areas of well-watered land. But there is no doubt that much of its available water occurs 'at the wrong place, or the wrong time or with the wrong quality' (Pigram 1988, p. 151). For example, about two-thirds of our runoff occurs north of the Tropic of Capricorn, in a zone that holds only about 5% of our population and little of our productive land (McMahon et al. 1992). Furthermore, economic hardship is caused not simply by aridity, but by variability. Lengthy and widespread droughts have been common since settlement. Australian streams are among the most variable in the world; and unlike streams elsewhere, those with the largest catchments tend to be the most variable. Not surprisingly, therefore, the volume of water stored per person is higher than for other countries (McMahon et al. 1992).

Figure 2.2 illustrates the disparity of runoff and available water across the continent: it is obvious that the pattern of settlement was not merely an accident of maritime discovery. Our highly urbanised society has become coast-bound not just because of settlement history, but also because of the pattern of availability of water. Hence, it is not surprising that many significant changes in the Australian landscape have been related to attempts to come to terms with the variable water supply. There have been extensive modifications to Australia's regional hydrology: the irrigation schemes in the Murray–Darling Basin and the

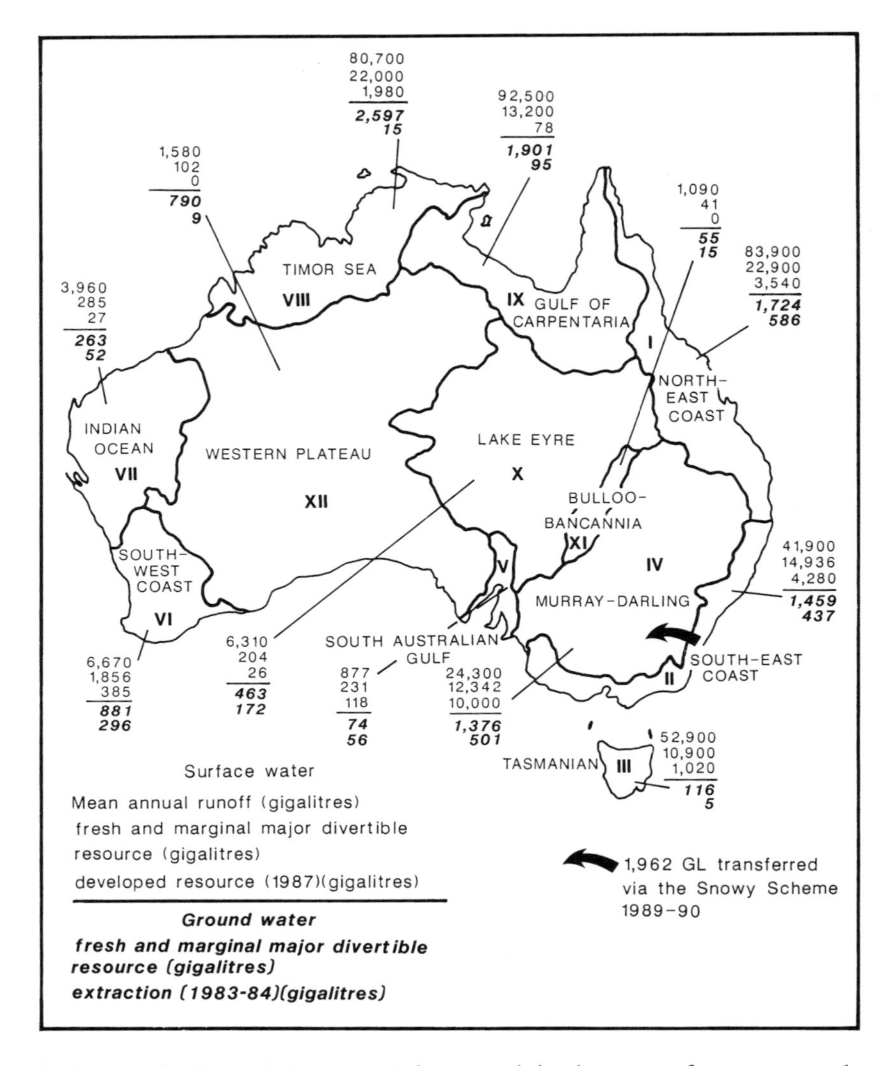

Figure 2.2 Drainage divisions, showing the available and developed surface and groundwater resources. (Source: data from ABS 1992.)

Ord River; the Snowy Mountains Scheme; and the diversion of water to supply cities such as Adelaide, Kalgoorlie and Broken Hill. Except during the Second World War, construction of large storage dams increased in all States throughout this century (Figure 2.3). Irrigation allowed more intensive agricultural development than dryland farming; large urban storages, especially around Sydney, allowed virtually unrestricted use of water by city dwellers; flood mitigation and lower impact of droughts justified large dams in agricultural areas; hydroelectricity was a spin-off on the mainland, but the main reason for the construction of dams in well-watered Tasmania. Until the Gordon-below-Franklin controversy in 1983, building large dams was an almost universally popular practice. This was despite the misgivings of commentators such as Bruce Davidson (see Chapter 1), who argued that irrigation was a result of poor perception, not a practical response to reality. After 1983 dam construction, either for industry or for irrigation, was far less acceptable to many people.

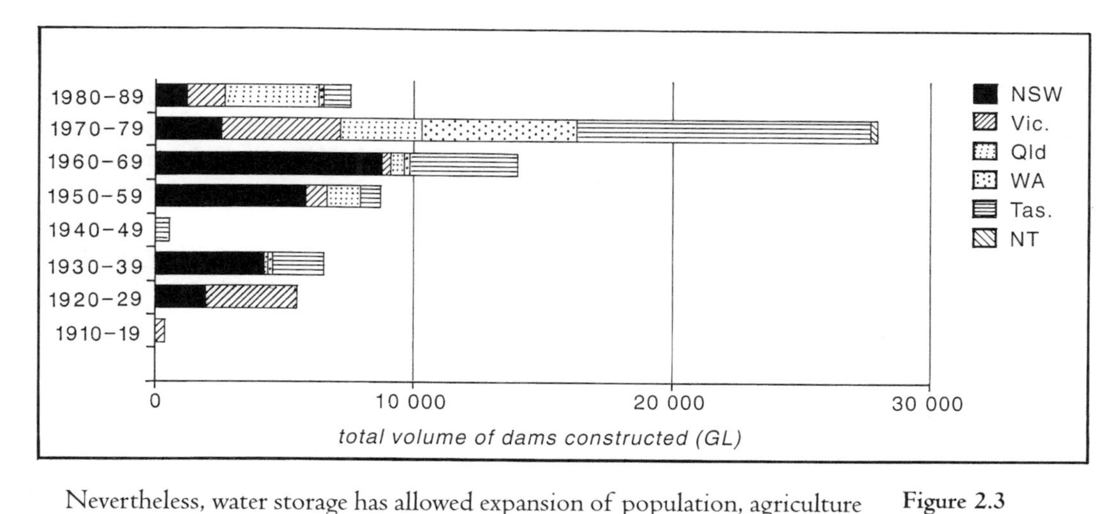

Figure 2.3
Bar graph showing volume of large dam construction by State since 1910. (Source: data from ABS 1992.)

Nevertheless, water storage has allowed expansion of population, agriculture and industry, and resultant changes in land use patterns. It has allowed modification of perhaps the most significant land-use change since settlement: clearing for agricultural development.

Measuring the changes: How much has been cleared?

As agriculture expanded, wooded lands were cleared (Figure 2.4). In Australia, as is the case globally, much of the publicity and popular sentiment is focused on the deforestation of rainforest and tall forests, but the extent and environmental consequences of clearing have been greatest in woodlands and more open country. Open forest, woodlands and open woodlands have been used for grazing. In the tropics the vegetation structure has not always been altered, but the floristics of the grassy understorey have changed in response to grazing and frequent burning. In the open forests woody shrubs have reduced and grasses

Figure 2.4
A patch of remnant forest on a cleared ridge on the Atherton Tableland, Queensland

increased; in the woodlands, grasses of *Themeda* genus have reduced and those of the less palatable *Heteropogon* genus (spear grass) have increased. In temperate areas the open forests, woodlands and open woodlands have been more extensively cleared, and often completely replaced by wheat or sown pastures of exotic grasses and legumes (Carnahan 1986). M. Williams condemns the clearing in trenchant terms:

> the woods, like other elements of the landscape, such as grasses, soils and waters (and even the inhabitants), were swept away in the pursuit of improvement, development and progress as then understood . . . The prodigality was matched only by the ignorance and misperception of the extent of the wooded endowment of the continent.
>
> Williams 1988, p. 115

Changes were rapid and irreversible. The decline and extinction of species occurred within decades; some lands were cleared and have never undergone a detailed survey, even of the remaining natural vegetation; and changes begun may still not be apparent in ecosystems where the lifespans of species are comparable to the period since settlement (Beattie et al. 1992). In the Gippsland region of Victoria, very little of the native vegetation survives because large areas have been cleared for irrigation and other farming. It would be a mistake, however, to see lack of trees as evidence of wholesale clearing. Lunt (1997 a and b) showed that many areas were naturally treeless grasslands (more because of soil conditions than burning by Aborigines). Where woodland was present, tree densities were low—from 5–9 trees/ha on land described by settlers moving in in the 1840s as 'open plains, good agricultural land' and up to 59 trees/ha on well-watered places. In contrast, tree densities in reserves and roadside verges now are much higher—up to 285 trees/ha. This means that we could look at these remnants of woodland vegetation and gain the impression of much denser tree cover in the past than is justified. Remnant vegetation patches are not pristine but are derived communities. Even if uncleared, they can be affected by factors such as changed fire regimes, grazing and rabbit predation, and—as Morcom and Westbrooke (1998) commented concerning the Wimmera—restorative tree planting could lead to communities becoming less, rather than more, like their former composition and structure.

Moreover, there are still large areas of woodland and forest, particularly along the eastern coast, and it is misleading to imply that all wooded country has been denuded (Plate C). Indeed, even in our metropolitan cities the number of trees is still great, and the 1988 vegetation map is correct in showing woodland over the coastal strip that includes Brisbane, Newcastle, Sydney, Wollongong and Melbourne.

The general picture—that clearing of wooded areas has been dramatic—is clear enough, but the detailed view is not well defined for the whole continent. In the 1992 Australian Bureau of Statistics publication on the Australian environment, a map of Australia's vegetation in the 1780s shows 'shrubland' along the Murchison River in Western Australia and on the Eyre Peninsula in South Australia (see Plate C). Yet in an earlier ABS publication—the *Year Book Australia*

1990—there was a map indicating that both these areas included forests or woodlands that had been grossly disturbed since European settlement.

The successor to the ABS (1992) *Australia's Environment: Issues and Facts* is the *Australia State of the Environment 1996* report (Department of Environment Sport and Territories 1996). In some cases the data shown on continent-wide maps is clearly a matter of interpretation rather than observation. For example, the comparative maps of 1788 (natural) and 1988 (present) vegetation show the same change across large areas—most of the Nullarbor, west of Lake Eyre, around Lake Torrens, and from the Flinders Ranges towards the Darling River. The change is not great—from low shrubland to low open shrubland, but it is obvious because the darkness of colours marking the two types of vegetation is very different. Since the boundaries on the two maps correspond exactly, it seems probable that it has been found that grazing in the arid rangelands has made the vegetation sparser within some of these areas, and that this finding has been extrapolated across all the areas. It does not mean that the change has been confirmed by extensive research.

Historical reconstruction of past vegetation is particularly difficult because information may be sparse and because it is hard to see past events through the eyes of past generations:

> [Clearing] was part of the normal farm-making operations and, as such, went unrecorded . . . Additionally one must remember that one is looking for something that no longer exists; it requires an enormous effort of the imagination to see the land as it once was.
>
> Williams 1988, p. 116

To quantify the change, Graetz et al. (1995) compared a satellite image of the continent captured in 1990 with the 1788 vegetation reconstruction map produced by AUSLIG (see Plate C). They defined two zones—an Intensive Land Use Zone (ILZ), occupying the 39% of the continent that holds over 90% of our population, and a more arid Extensive Land Use Zone (ELZ), occupying 61% of the continent but being sparsely populated. Forests and woodlands on the 1788 map were identified and compared with the existing vegetation as shown on the 1990 satellite image. They found that 1.03 million km^2 or 50% of the ILZ had been cleared or thinned, and 1.13 km^2 or 24% of the ELZ woodlands had been significantly or substantially disturbed. Even with this careful and sophisticated estimate, the authors sounded strong notes of caution. The resolution of the 1788 data was at best 1 km^2 but more realistically 25 km^2, whereas the satellite image pixel size was approximately 1 ha (0.01 km^2) re-sampled to 1 km^2. Also, all height categories and overstorey genera categories from the 1788 data were generalised. Thus, the two sets of data being compared were dissimilar in scale, generalised and, in the case of the 1788 data, derived rather than observed. They are useful, but as best estimates, rather than as hard facts.

Other inconsistencies in the assessment of clearing rates seem almost certainly to be due to altered definitions. Between 1980 and 1990 the area of rainforest in New South Wales declined from 300 000 ha to 265 000 ha (ABS

1992). Given the controversies over logging and the declaration of the Australian East Coast Temperate and Sub-tropical Rainforest Parks in 1986, this seems reasonable. There has been a loss of resource, but the greater protection now may eliminate further loss. Yet in Queensland and in Tasmania the same source of data shows substantial increases in the areas of rainforest. Rainforest areas increased in Queensland and in Tasmania in the ten years from 1980 to 1990. Both these States experienced controversies over rainforest preservation, which also ended in the declaration of World Heritage areas (see Figure 1.7). Why did their areas increase when the area in New South Wales decreased?

Areas of rainforest recorded in *Year Book Australia* from 1980 onwards show that the picture is confusing. Between 1980 and 1997, the area in New South Wales has decreased, increased and then decreased again, while in Tasmania it has increased, then decreased. The area in Tasmania rose sharply from 499 000 ha in 1988 to its 1990 level of 605 000 ha (Table 2.2). There seems little doubt that this figure rose not because of real increase in the area of rainforest over a mere two years, but because of reclassification of 'not-rainforest' areas. Even more dramatic and less convincing is the reported doubling of the area in Queensland between 1993 and 1997! The point to note is that we might have accepted without question a decrease in area—for example, the summary figure showing a decline in New South Wales between 1980 and 1990—as being a real change. But we need to consider the possibility that decreases, as well as increases, could be a matter of reclassification. In short, we need to be wary of uncritically accepting figures that support our preconceptions about environmental change.

Table 2.2
Areas of rainforest (thousands of ha) in several States, 1980–97

Year	*NSW*	*Queensland*	*Tasmania*
1980	300	1074	472
1982	253	1074	472
1984	253	1074	499
1986	265	1237	499
1990	265	1237	605
1993	260	1237	565
1997	209	2567	545

Data from ABS 1992, and ABS 1982–99

Changing perceptions: What is a resource?

Many major environmental controversies arise from varying or conflicting views about the value of the resources in an area. The idea of conserving the Kakadu area had already been floated before the uranium resources of the Northern Territory were confirmed in 1970 (in response to exploration encouraged by the Commonwealth government). Proposals for a national park had first been made in 1965 and the idea was given greater urgency by the proposals for uranium mining at the Ranger site at Jabiru (Ranger Uranium Environmental Inquiry 1976, 1977). The boundaries of Kakadu National Park were gazetted in 1975, but it is easy to see that the conflict about the proper use of the area is unresolved. Public concern continues but on varying grounds—the Coronation Hill controversy, issues of Aboriginal land rights and title, resurgence of concern about the development of nuclear facilities in Australia, approvals for expansion of uranium mining. All illustrate the ongoing and ever-changing nature of the issue, but perhaps the most telling illustration of the variability of the perceived importance of issues is that the *Australia: State of the Environment 1996* report does not list Kakadu or uranium in its index.

Changing understanding of resources can lead to changes in land use. For example, the Hunter Valley of New South Wales has been a major coalfield since the mid-1800s, and the steelworks complex at Newcastle was established because of the area's coal resources. Until the 1970s, however, coal mining was mainly carried out by underground methods and confined to the lower parts of the valley, from Cessnock eastwards. The upper Hunter around Muswellbrook was an agricultural area, with dairy and beef cattle. Changing technology—particularly the development of open-cut mining—and competition with the open-cut mines in Queensland led to the recognition that the upper Hunter area was valuable for coal extraction. There, an anticlinal structure brought the coal seams close to the surface, and the exploitation of these has led to the development of major mines for export, as well as for local generation of electricity. Large dam construction and interbasin water transfers have accompanied these changes (Figure 2.5).

As the resources have been exploited, the controversial environmental issues have also changed. In 1980 the Hunter Valley was a rapidly industrialising landscape. Two new power stations, each of 2640 megawatt capacity, were being

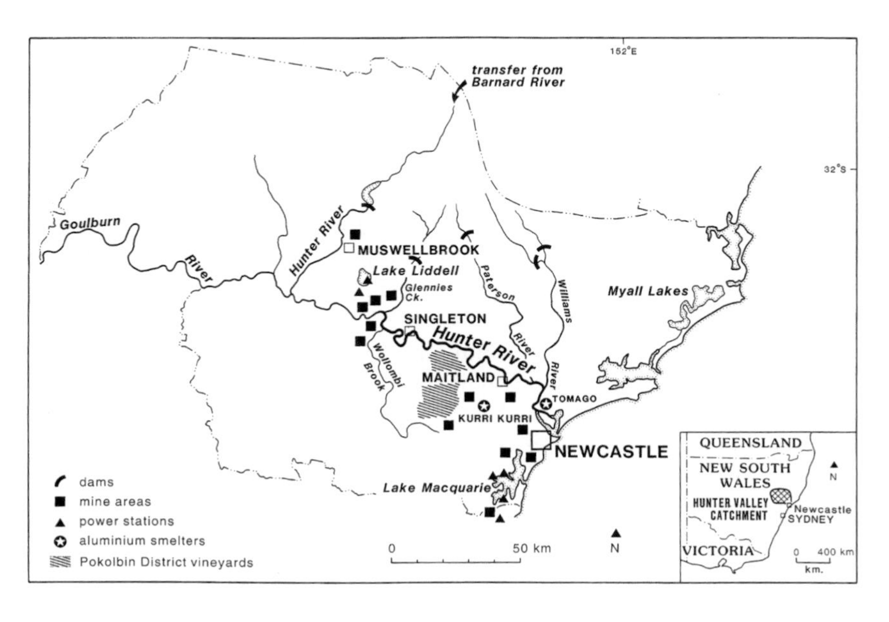

Figure 2.5
The Hunter Valley region of New South Wales and its major resource developments.

built near Muswellbrook; existing coal mines were planning to expand, and more than 20 new mines were proposed; expansion of an existing smelter and construction of two new smelters would increase aluminium production eight-fold (State Pollution Control Commission 1980). These were interdependent industries. Coal mined would fuel the power stations, which would supply the smelters. Air, water and noise pollution from all industries were recognised problems, but the most controversial issue was the emission of fluorides from the smelters. Fluorides can directly damage the leaves and fruits of many plants, and interfere with plant growth and reproduction. Grape vines are particularly susceptible, and two of the smelters were very close to the important Pokolbin District vineyards. The Hunter Valley is one of Australia's longest-established and largest wine-producing areas. A threat to its vineyards could have consequences more significant than mere economic loss, and stringent control and monitoring programs were put into place.

Drought from 1979 to 1983 changed the focus of controversy. The new industries, particularly power-generation and washing of coal, competed directly with irrigation agriculture for water. They added a much higher value per kilolitre of water used. They encouraged urban development, which also competed for water supplies. Generation of power was seen to be restricted by scarcity of assured water supply, and pressure for new storages and for transfer from the Barnard River increased. Temporary water restrictions and cessation of the issuing of new irrigation licences was necessary, but the underlying question was one of policy: should allocation be biased towards traditional agricultural uses, or modified to achieve greater economic efficiency? This is a complex issue of regional planning, further complicated by controversy about water quality (Day 1986).

The open-cut mines intercept saline groundwater, which must be disposed of on-site or released into the Hunter River, generally during times of high discharge so that the salinity is diluted. Farmers downstream argued that this increased the salinity of the water that they required for irrigation. Mining companies replied that irrigation was the more significant cause of salinisation of the river. As the drought broke and the coal industry faced economic difficulties, the issue faded, but it has by no means disappeared. By the end of the 1980s controversy centred on air pollution again, but this time in relation to emissions from the power stations and their effects on human health (Henry et al. 1989). There was found to be a higher incidence of chronic asthma close to the Munmorah Power Station than in the control area (Nelson Bay), although the air quality—as indicated by levels of sulphur and nitrogen oxides—were within recommended guidelines. In the early 1990s, following trends elsewhere in Australia, perhaps the most controversial issue in the Hunter region concerned the old industrial area of Boolaroo, and the health impacts of emissions from the lead smelter there. Today the competition for water supply has re-emerged as the government sets environmental flow targets, designed to reduce or repair the impacts of river regulation on ecosystems. These will alter the allocation and thence cost of water to the power generation and irrigation industries.

The debate about environmental quality in any region is dynamic. As this brief review of the Hunter Valley illustrates, the debate's focus changes as:

- a new resource is recognised and developed
- natural fluctuations, such as drought, place pressure on shared resources
- potential problems are seen to be controlled by technology or management
- economic conditions alter
- broader trends and fashions in environmental issues are followed.

Monitoring changes: How severe is erosion?

Many writers have discussed an increasing environmental awareness in the last few decades, both in Australia and worldwide. This altered awareness has been accompanied by a changing mood among Australians: from one of optimism and regard for 'progress', to one of pessimism and the conviction that most change has meant environmental degradation. The mood is captured well by the poet and environmentalist Judith Wright in her poem 'Eroded Hills':

> These hills my father's father stripped
> and beggars to the winter wind
> they crouch like shoulders naked and whipped—
> humble, abandoned, out of mind.

As with the issue of clearing, it is easy to make a generalisation—erosion and land degradation has increased since European settlement. It is more difficult to say exactly where and by how much. For a start, data may not be available. For water resources, a matter to which Australians have paid a lot of attention, only about 50% of the total yield is measured. Data for even this are often inadequate in terms of length of record. Most of the measurement takes place in the well-populated areas where major water developments are located, and the errors of estimation elsewhere may be as high as 30% (Warner 1986). Even for this crucial and well-recognised aspect of the landscape, the available data are inadequate. When we come to look at erosion, the data are sparser still. R. C. Burgess et al. (1989) found it hard to distinguish between the relative contributions of natural and anthropogenic erosion to total wind erosion. In some cases, such as the Western Australian wheat belt, erosion due to human activity did not appear to be high. This might have been due to inappropriately located measuring stations, since there is other evidence for considerable wind erosion of the cleared wheat fields.

Monitoring the changes that have been made has become increasingly important, but it is not easy to find the information that is needed to do this effectively. Even when we have information, we have to make generalisations in regard to large areas. To assess the extent of land degradation in New South Wales, aerial photographs were used, but clearly it would have been too time-consuming to look at photographs of the entire State (Graham 1989). Sample areas were selected at 5 and 10 km intervals using map grids, and the extent of land degradation within a 100 ha circle of these points was assessed from aerial photographs. The severity of degradation in that circle was then assumed to apply to the whole 50 km^2 or 100 km^2 area. Hence, the final map showed which of those 50 km^2 or 100 km^2 areas contained land degradation problems of particular types (acidified soils, salinity, gullying, sheet erosion, wind erosion, woody shrub infestation). They do not show that the whole of each area was affected, but this can be the impression they give. The maps seem to overstate the severity of problems by making land degradation appear very widespread, but this is the consequence of the generalised nature of the data.

It is not surprising, therefore, that data about the Australian environment are often difficult to interpret, and may seem contradictory. The Australian Bureau of Statistics publication *Australia's Environment: Issues and Facts* (1992)—which summarised and explained statistics about environmental issues, and brought together material from many governmental and academic publications—was an important initiative. But it contained some inconsistencies that highlight the difficulty of obtaining precise environmental data. A map of 'areas susceptible to water erosion' presented a vivid picture of a widespread problem. The caption was puzzling; it described five classes of 'cropping areas subject to . . . damage', ranging from 'nil to low' up to 'regular' and 'severe' categories. This implied that the water erosion was linked to the presence of cropping land, yet many of the areas shown lay outside the zones of 'pasture and cropping' provided on the map of 'present vegetation' only a few pages before (Figure 2.6).

This discrepancy can be explained. In assessing the degradation for all regions, an estimate was given of the severity of erosion in cropping lands. The assessment took account of the regions outside the main cropping areas (those shown on the land-use map), even if the extent of cropping in that region was very small (Rose 1993). Hence there are very small patches susceptible to erosion within the very large areas mapped as being subject to some damage, but

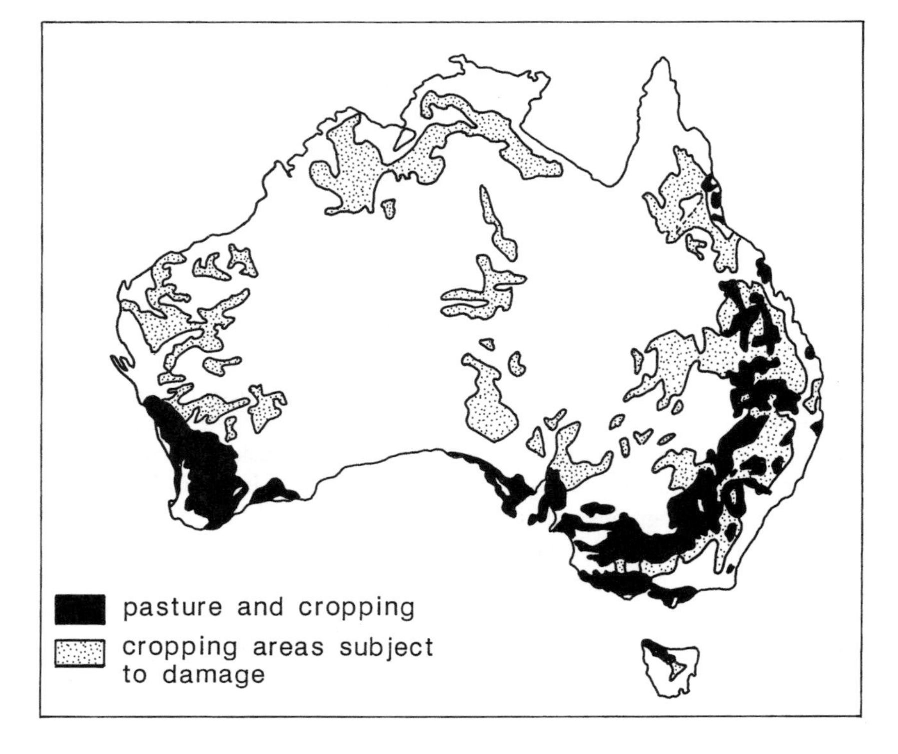

Figure 2.6
Pasture and cropping lands in Australia, and the areas outside those lands that have been categorised as 'cropping areas subject to damage'. (Source: data from ABS 1992.)

lying beyond the main zone of cropping and pasture. The general impression conveyed by the map overstates the damage.

Finding the real culprit: Natural and anthropogenic change

We must also be wary of attributing degradation entirely to human activity—particularly since European settlement—where natural causes may be significant. In some places the role of natural changes has been overlooked. The great drought of the 1930s caused severe wind erosion over much of south-eastern Australia, and severe dust storms resulted. These were particularly severe in the semi-arid parts of South Australia:

> I suppose you remember those days when, heralded by brownish red clouds on the horizon at first no bigger than a man's hand like the one in the Bible, Peterborough turned on one of its spectacular dust storms which enveloped the town in clouds of whirling red dust making it necessary to have the electric light on in the house even at high noon. The dust always got in, necessitating an entire house cleaning after the storm had passed away.
>
> Jefferey Johnson 1993, letter to Peter Johnson, in author's possession

> Sometimes the storm would come as a dirty, eyefilling, itchy-arm-making half gale. The one I remember came silently, just rolling over everything in its path. I looked up at the sun—and it was a dull, blue disc. At this time of year, nobody could stare at the midday sun without being blinded but on this occasion, the fiery sun was a blue moon!
>
> Peter Johnson 1994, letter to Jefferey Johnson, in author's possession

At the time of the dust storms some writers blamed ploughing and loss of soil humus, while others pointed to overgrazing by stock and rabbits. More recently an extraordinary change in zonal winds during that decade has been identified. While the changes due to cropping and grazing undoubtedly worsened the effects, the two major causes of wind erosion were natural: drought and an unusual wind pattern (McTainsh and Leys 1993).

Erosion and gullying have certainly followed clearing and grazing. Studies of cores from a lagoon near Glen Innes, New South Wales, showed changes in sediments and pollens, and increases in phosphorus levels first in sediments at 1 m dated by ^{210}Pb methods at 1813+/–12 and 1836+/–7, and then dramatically at 0.2 m dated at 1960+/–1 and 1969+/–1. These dates correspond to the known time of settlement of the area (1837) and to the expansion of the use of tractors and superphosphate after the 1950s (Gale et al. 1995). Yet we need to remember that gullying is a natural feature of much of the Australian landscape. In south-eastern Australia, widespread land instability occurred in the cold dry climates of the Last Glacial period, 25 000 to 15 000 years ago. Aborigines had been occupying the area for longer than this, at least since 30 000 to 40 000 years ago, so that burning practices would have been well established before the erosion during the Last Glacial. In the past, climatic change has caused major increases in erosion that are independent of the actions of people. The gullying that concerns us now is often cutting back through deep, old alluvial fills. These could have formed only as a result of severe and prolonged erosion of the surrounding hillsides well before European settlement. But more importantly, they often show evidence of multiple phases of gullying as well (Box 1).

A study of a typical catchment in the Southern Tablelands of New South Wales revealed that deep gullies had affected parts of this catchment long before settlement, eroding and then aggrading again over a period spanning at least 8000 years (Prosser 1991). But after settlement in the 1830s the gullying increased rapidly. No longer were gullies scattered over the catchment; within 70 years they were ubiquitous. They were initiated by localised disturbances such as roads or drains, and extended rapidly up the creeks. Do we now find a devastated landscape in the catchment? Has the early rate of erosion been maintained? The answer to both questions is 'no'. In fact, most of the gully networks in this and other catchments of the region were formed last century and were stable by 1945. The gullies are now large enough to carry any flows, even in high rainfalls, and the subsurface water is not causing appreciable piping and headward extension although the flow from them is turbid. The degradation, although severe initially, has not continued indefinitely. This adds a complicating factor to attempts to predict future environmental degradation: the rate of change, particularly soon after that change is initiated, may be rapid, but it may not be maintained. Indeed, as the survival of large volumes of Pleistocene as well as Holocene sediments in valleys on the tablelands (Box 1) and in the coastal valleys of south-eastern Australia demonstrate, the volumes of sediment moved and changes of channel over the past 200 years are of much lower magnitude than the changes prior to settlement.

Box 1
Erosion at Limekiln Creek, Bungonia, New South Wales

Dramatic gullying is usually attributed to changes that have taken place since European settlement of our agricultural areas: trees have been removed; dryland salinisation has made the soils more erodible; grazing by stock has left the ground bare and compacted. All these changes can lead to erosion and gullying. However, this may not be the full story, and there are several questions we should investigate:

- Is the gully eroding down into weathered rock, or is it cutting through old sediments?
- Is the present erosion more severe than any in the past?
- Is there clear evidence that erosion began, or at least accelerated, after settlement?

Detailed work at a small creek—Limekiln Creek, near Bungonia, 30 km east of Goulburn in New South Wales—helps us to answer these questions.

Figure 2.7
Severe gully erosion, with soil pillars undercut by subsurface piping, near Bungonia, New South Wales.

The gullying shown in Figure 2.7 has eroded some weathered bedrock, but nearly all of the erosion has occurred in old sediments. The present gullying, severe though it is, has not cut down as far as the floor of the original valley. At some time in the past, the stream cut into the Ordovician metamorphic rocks that underlie the gullied area and crop out on surrounding hills; nowadays, throughout its length, the present creek flows through sediments that infill the former bedrock valley. These old sediments infilling the valley are evidence of gully activity in the catchment, which took place long before the land was cleared for grazing. There has also been severe hill-slope erosion in the past. Alluvial fans of coarse angular debris blanket the lower slopes beside the creek to depths of over three metres. Thus, the present gullies are excavating a complex sequence of Quaternary sediments; these sediments are themselves evidence of severe erosion in the past. Reconstructing the sequence of deposition, and looking at the characteristics of the sediments, throws light on the present processes.

Basalt flows, whose remnants still reach close to the modern valley, flowed from vents and into the creek about 46 million years ago. One of these flows dammed one arm of the creek, and there are still remnants of the laminated sandy clay deposits from small temporary lakes that were impounded behind the basalt dam. However, the creek gradually cut through the dam, and most of the lake clays were washed downstream and redeposited. Weathering over a long period caused mottling of the redeposited clays, and these mottled clays can be found near the present base of the creek along most of its length. Above these mottled clays, in some places, there are alluvial deposits of orange sands. These were deposited when iron-rich sands were eroded from the surrounding hills and washed into the creek. Thermoluminescence-dating of the orange sands shows this occurred between 56 000 and 38 000 years ago. At that time these sands were trenched and overlain by black clays, as the basalt weathered and was eroded from several places in the catchment. This erosion of the weathered basalt occurred more recently than 27 000 years ago. Uranium–thorium dating has been used to obtain an age for manganese-rich crusts that had developed on the orange sands and on fan deposits; the crusts are younger than the orange sands, but older than the black clays. In the black clays there are numerous nodules of calcrete, indicating a period of stability and dry conditions; a radiocarbon age of 1300 years was obtained for charcoal in these clays. Thus, the pre-European landscape was often unstable and being eroded (Figure 2.8).

At present the pattern of the erosion depends on the sediment through which the gully is cutting. In the downstream reaches, where the creek is cutting mainly through the orange sands and overlying black clays, the trench is large enough to carry flows at high water levels. The spectacular erosion is confined to one arm of the creek, where the gullying is excavating old lake clays, deeply weathered bedrock and alluvial sandy clays. If we take this zone of rapid erosion, use aerial photographs to estimate its rate and then calculate how long the whole gully has taken to form, we find that it has taken 600 years. Therefore, it cannot just be a result of clearing since European settlement.

Prior to European settlement there was probably an open eucalypt woodland over most of the catchment, although it is possible that the valley floors were swampy and sedge-covered. Clearing may have reduced the amount of water removed from the soil by reducing evapotranspiration. Certainly the A horizon of the soil, even beside the deepest parts of the gully, now stores enough water to remain soggy, even in dry weather when the gully floor has minimal flow.

Figure 2.8 Evidence of severe pre-settlement erosion near the Bungonia gully. During the past 27 000 years black clays have infilled channels (to the level of the man's shoulder). These channels are cut into older orange sands (exposed on the floor of the present gully), which are themselves infilling a deeper and older channel.

Thus, clearing may have accelerated the process by increasing available soil water near the top of the B horizon. But three facts are clear:

- Severe gullying occurred in the past, and the stream has been infilling and then re-eroding its course for at least 38 000 years. There have been periods of stability and periods of erosion throughout this time.
- The present dramatic gullying may have been accelerated by modern changes such as clearing, but almost certainly began well before European settlement of the area.
- The pattern and severity of gullying is determined mainly by the character of the sediments that the gully reaches and cuts through.

The site provides an excellent example of the need to be cautious in attributing all land degradation to post-settlement changes.

Details of dates and stratigraphy are from Wray et al. 1993

When dealing with entire catchments, separating anthropogenic and natural changes is problematic. Young et al. (1986) raised the question of whether clusters of similar radiocarbon ages in sediments provided evidence for climatic changes over the past 40 000 years in the alluvial sequences of coastal New South Wales. They pointed out that re-working and destruction by weathering affected the apparent patterns of dates, and that clusters of young dates were at least as likely to be due to random erosional events that triggered switches across geomorphic thresholds, as to regional climatic changes. And the scarcity of dates from the last 200 years raised doubts about the impacts of burning or land clearing. In contrast, Page and Carden (1998) argued that clearing and grazing in the Tarcutta Creek catchment of southern New South Wales led to increased flooding and sediment supply, and thence to a switch in channel form from a stable pattern of deep pools and extensive swamps, to a sandy stream with alternating reaches of erosion and deposition. In short, they saw the switch in channel pattern as the crossing of the intrinsic threshold between meandering and braided stream form but as due to human-induced changes after clearing.

Even with flood records within historical times, it has proved difficult to separate the roles of climate and human activity in altering channel form. Some researchers (Erskine and Warner 1988, Nanson and Erskine 1988) have argued that climatic changes since settlement have caused catchments in south-eastern Australia to be shaped alternately by flood- and drought-dominated regimes. In flood-dominated regimes, the channels become wider and shallower; in drought-dominated regimes, they become narrower and deeper. These climatically driven changes mask or outweigh any anthropogenic changes. In the view of Kirkup et al. (1998, p. 242), this theory 'has seriously underplayed the significance of European disturbance to river channels and catchments'. They dispute the validity of the flood records used to derive the theory, and the statistical methods used to interpret these. They do not dispute that there has been one significant climatic shift to a higher rainfall regime since the late 1940s, but do not accept the theory of multiple switches in regime. In reply, Erskine and Warner (1998) reassert their position by presenting further discussion of statistical methods. The aim here is not to decide between the competing views, but to point out that both are derived from essentially the same data sets but are arrived at by differing methods of analysis.

Problems of scale: How precisely can we measure?

That the result of a measurement depends on how the measurement is made, and also on the scale of the item measured, is a truism but is often neglected. If we change the way in which we measure something, then we can change both the results which we get and their accuracy, so that we may be unable to tell whether the property measured has changed or not. Consider water quality testing. When attempting to analyse monitoring results, it is often frustrating to find that the sampling site was changed, that the sample was taken by grab sample sometimes and then by bulked samples later, that some results are from field testing kits with limited accuracy and others by one or more laboratory methods which may be more precise but may be affected by changes in the sample during transport to the laboratory. These changes in the way the water quality was measured affect the validity of any analysis of the results. Also, the scale at which we measure must be taken into account. Everything we measure is spatially contingent and we cannot extrapolate simply to larger or smaller scales (Box 2). Both method and scale may vary for different purposes, and all these variations can complicate the comparison of data for different areas or different periods.

Box 2
Scale and measurement

- It is easy to be confused about the terms 'large scale' and 'small scale' maps. We feel that we should see more on a 'large scale' map, and this is in fact so, but when we look at the scale written on the map, confusion can arise. For example, a 1:100 000 map (a standard scale for Australian topographic sheets) shows far more detail than a 1:2 million sheet (which could show all of Tasmania on an A4-sized piece of paper), yet 2 million is a larger number than 100 000. The answer is simple. The scale we write as 1: (number) is really a ratio, called the representative fraction or RF. One unit on the map represents that number of units on the ground. Expressed as a fraction it is easy to see that 1/100 000 is a larger number than 1/2 000 000.
- On a 1:100 000 map, 1 cm on the map represents 100 000 cm or 1 km on the ground. On a 1:2 million map, 1 cm represents 20 km. This has implications when we measure distances using maps. We can measure with a ruler to about 0.1 cm (1 mm). This means that a thin line such as a stream course or vegetation boundary on our map will represent a thickness of 100 m on a 1:100 000 sheet or 2 km on a 1:2 million sheet. Obviously this does not mean that the stream or boundary actually is that width; it is just the way it is marked on the map. It also means that any measurement we make from the map is only accurate to this distance. So, if we measure a distance across the map of 1 cm (to estimate the width of a particular feature, for example), then there is an inherent measurement error of +/– 0.1 cm or +/– 10%. If we measure over a longer distance the percentage error is smaller, but we need to keep it in mind.
- In these days of photocopiers we need to be particularly careful. If we copy a map, it is likely to be distorted at the edges. Also if we enlarge or reduce it, then the RF is no longer valid. It has changed by the factor of reduction or enlargement. For this reason, it is a good idea to draw linear scales on diagrams because these keep constant proportion when enlarged or reduced.
- Especially when dealing with maps of large areas, we need to be aware too of the role of map projections. Because maps are two-dimensional planar representations of a three-dimensional curved surface, they are not true in both scale and direction except over small areas where errors due to the distortion of the earth's surface are insignificant. The clearest example is the Mercator map of the world, commonly shown on children's desks. This projection is a form of cylindrical projection, where the lines of latitude and longitude are projected onto a cylinder which touches the globe along the equator. Hence the lines of latitude near the poles, which on the globe are much shorter than the equator, are shown as the same length as the equator. Countries in high latitudes such as Greenland seem much larger than they really are.
- Caution in dealing with scale and measurement is needed in the field too. Data collected at one scale cannot be simply multiplied to estimate results at a different scale. If we try to do this, we assume that the relationships we are investigating do not vary from scale to scale. The fact of the matter is that

most natural characteristics, not only of mobile animals but also of in-situ properties like soil type, vary greatly. At a small scale, the pH of soils is usually lower in humid areas than in arid areas, but at a larger scale it will vary with parent material, vegetation, slope position, water flow, land use and other factors. As these factors vary across different scales, so their influence will affect pH across different scales.

This problem can be illustrated by an apparently simple question: how long is Australia's coastline? Official estimates available in the 1970s ranged from about 19 000 km to over 36 000 km; another estimate yielded 132 000 km, which is more than three times the circumference of the earth. Researchers with the CSIRO (Galloway and Bahr 1979) used three methods to measure some lengths of coastline from 1:250 000 scale topographic maps. They used dividers to step out the lengths, a measuring wheel (opisometer) and fine wire to follow the outline of the coasts. The opisometer method was inaccurate; the dividers and fine wire gave consistent results. Again using the 1:250 000 scale maps, for the whole of Australia this time, they found that the coastline was 50 020 km long if they measured along the coast with a divider spacing of 0.5 km. Measuring with a divider spacing of only 0.1 km allowed them to include the lengths of smaller embayments and irregularities along the coast. This time, the total length measured was 61 600 km. But if maps at a coarse scale of 1:2 500 000 were used with a wide divider spacing of 500 km, this very rough estimate gave a length of only 12 000 km. If even less accurate maps, at 1:15 000 000 were used, the result dropped to 11 800 km. The scale of measurement changes the result dramatically.

This is not a trivial matter. As R. W. Galloway and M. E. Bahr emphasise, the scale of measurement needs to be tailored to the information needed. If the intention is to estimate the proportion of scenically important cliffline along a particular stretch of coastline, then 500 m could be an appropriate unit to measure the length of the coast. If the aim is to estimate how many building sites would be available for urban development, the appropriate unit might be 25 m. But if the project is to estimate the population of intertidal molluscs along the same stretch of coast, a more appropriate unit may be only 1 m. Taking these three units for one stretch of coastline yielded lengths that varied from 8.2 km to 16.8 km. Thus, there is no one 'correct' answer. The smaller the scale of measurement, and the more accurate the map, the longer the coastline. This is one example of a very important general principle. Measurement of environmental 'facts' gives results that depend on the method and scale of measurement. Unless both of these factors are the same for different sets of measurement, we cannot make direct and valid comparisons.

As we noted earlier, it is very difficult to obtain comparable data and then to make assessments that are unambiguous, particularly where assessments of very large areas are involved. The early assessment of land degradation by L. E. Woods (1984) is still the most quoted national view, and the one on which most estimates of the national significance of problems are based. It is the summary maps that are frequently quoted, and it is easy to forget the warnings in the introduction to Woods's report. Woods comments that there are inevitable inconsistencies, as different soil conservation authorities approach problems differently. Estimates of land degradation over the whole continent were made by sampling areas of a 2 km radius. If any part of the sample area was degraded, the whole of the area it represented was shown to be in need of treatment measures; about half of the farm land was shown in this way. In 1990 the Commission for the Future took this to mean that half of the farm land in Australia was degraded—a much-exaggerated version of the original assessment. As Barr and Cary (1992, p. 283) comment, 'from this statement it is but a small step to a popular vision of half of Australia's farm land being either erosion gullies or white salt pans'.

Conclusion

No one doubts that changes in the Australian landscape since European settlement have been widespread and damaging. The settlers had minimal understanding of the character of the landscape, coupled with a determination to force the land to yield a living for them. Now there is a strong sense of urgency in the bid to halt and reverse the damage, and develop sustainable land management. We have a far better scientific understanding now than previously, but we need to approach 'facts' with caution. We interpret the 'facts' differently as community perceptions change. We recognise new resources and may disparage older ones. In many situations we simply do not have enough data to distinguish clearly between natural and anthropogenic change. The data we have is collected and classified by different methods, at different scales and by people with different perspectives. We are not completely unable to judge the significance of changes, but we do need to guard against careless use of statistics on environmental issues.

Chapter 3

The Effects of Agriculture

Introduction

AGRICULTURE—including cropping and pastoralism—is the most extensive and varied land use in Australia (Figure 3.1). It occupies 60% of the nation's land and includes a wide range of activities, from intensive irrigated production of exotic crops on small holdings, to extensive grazing of native pasture on vast stations. Yet of the 77 million ha suitable for cropping and improved pasture, only about ten million ha are free of constraints such as erodibility or potential salinisation of soils (McTainsh and Boughton 1993). Drastic alterations to Australian vegetation and soils, and consequently to faunal habitats, have occurred because of clearing and the introduction of exotic species to enhance agricultural production. The damming of rivers, and associated widespread ecological changes, have been prompted mainly by the demand for water for irrigation. In the late eighteenth century developing agriculture was necessary for survival and for the establishment of export industries. Until the mid-twentieth

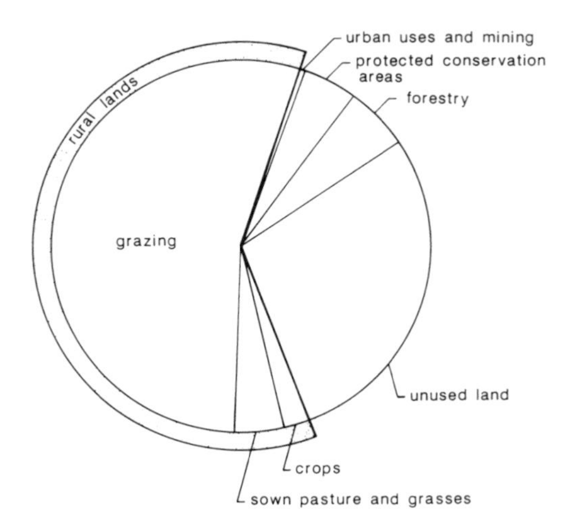

Figure 3.1
Pie chart of land use in Australia, showing the dominance of agricultural activities. (Source: data from ABS, *Year Books, Australia.*)

century it has provided the basis of our wealth and our expanding population. Yet, apart from some moves towards soil conservation, its environmental consequences have only very recently become a matter of strong public interest, not only in rural areas, but also within conservation movements. An increasingly condemnatory attitude towards rural land use has, perhaps only coincidentally, followed the increasing urbanisation of Australian society. More and more Australians are living in cities, particularly the capital cities; in 1921 the ratio of rural population to urban population was 38:62; by 1986 it was 15:85.

The first major development of agriculture came after the 1830s, as squatters moved out from the established settlements. This expansion into previously unsettled areas was anarchic and uncontrolled, and its impacts were devastating. By 1900 sheep and cattle had spread throughout the eastern States, over most of South Australia and the Northern Territory, across the eastern half of Tasmania, and about 500 km inland along the Western Australian coast from the Kimberleys to the edge of the Nullarbor Plain. The expansion of dryland cropping, predominantly of wheat, accompanied the spread of the railways, beginning in the 1880s (Jenkin 1986) (Figure 3.2). Many of the holdings were too small to be viable, and selectors were forced into an exploitative transitory cropping, which then led to soil exhaustion, erosion and retreat from marginal lands. Soldier settlement schemes after the First World War established more small holdings, which again proved inadequate in the Australian landscape. For example, in Victoria irrigation without adequate drainage led rapidly to salinisation; a drop in butter prices made the small dairy farms economically unviable, and many people were forced off their land. In the pastoral zone the 1930s and 1940s were the 'erosion decades', as over-use, drought and rabbits took their toll on the land (Barr and Cary 1992). Damage in South Australia was so severe that a 1942 book on the topic was titled *Australia's Dying Heart*. Nevertheless, this damage was only the final result of a process that began with settlement. It is likely that 70% of land degradation in semi-arid and arid regions occurred in the first 20 years after settlement (Noble and Tongway 1986a).

After the Second World War crop yields increased as erosion control began to take effect, new legumes were introduced and rotational cropping became

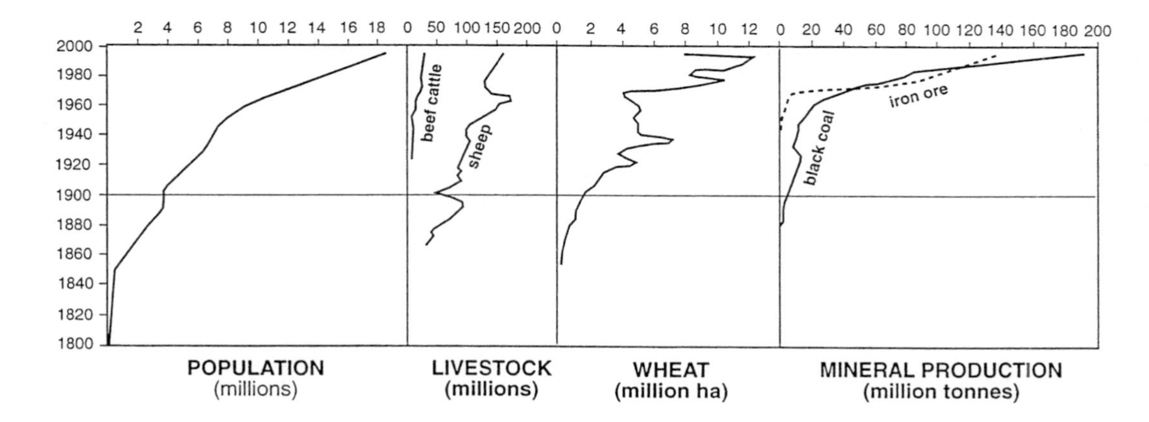

Figure 3.2 Graphs showing the expansion of agriculture and mining that has occurred with the growth in population since 1800. (Source: data from Jenkin 1986, and ABS, *Year Books, Australia*.)

popular. Agriculture expanded, but some adverse effects accompanied the industry's growth. Governments invested heavily in irrigation schemes, despite the known problems of salinisation. Tree-clearing gathered momentum, as cultivation replaced grazing land in many regions. Heavy tractors could work the black soil plains of northern New South Wales, so wheat was grown on former grazing land. Expansion of sown pastures led to the clearing of land and the intensifying of farming on hilly country in the Mallee of south-western Western Australia, and the brigalow country of south-western Queensland. There was an economic incentive to clear land, as the costs could be written off against taxable income from 1947 until 1983 (Barr and Cary 1992).

There is now some resistance to continued clearing, and a move towards re-establishment of trees on farms has taken place. The ten years from 1990 to 2000 were declared the Decade of Landcare, and there is strong government involvement in policies designed to reverse land degradation.

The magnitude of the problems

As discussed in Chapter 2, estimating the magnitude of problems such as land degradation is difficult. The most quoted study of the whole of Australia is that conducted by L. E. Woods (1984), which was based on reported assessments from soil conservation specialists. More recently the advent of satellite imagery has led to attempts to standardise analysis of change over time. An example is the published report of the Australian Case Study for the International Space Year by D. Graetz et al. (1992). This presents satellite images showing changes in land cover between 1972 and 1992. In some cases the changes shown are clearly discerned, and the accompanying text emphasises the dramatic nature of change. However, these case studies are still snapshots, chosen because of the significance of the areas (for example, the major cities) or because clearing or transition from extensive grazing to cultivation has been striking (Plate D). As Graetz and his co-authors contend in their introduction to the report, there is still no national assessment of the state of our renewable resources (that is, land cover and soils).

Australia: State of the Environment 1996 (Department of Environment Sport and Territories 1996, pp. 6–28) has a map of rill and sheet erosion (low, moderate and high categories). The caption is carefully worded: the map is 'an estimate . . . based on a soil erosion equation . . . [and] probably overestimates erosion in the seasonally wet tropics and in areas of rock outcrop'. The technical paper from which the map is taken (Rosewell 1997) has even more cautionary comments. The map was constructed by taking a soil erosion formula (the Universal Soil Loss Equation—see later in this chapter), then estimating the parameters for the formula from a digitised soil map (assigning a soil hydraulic conductivity value on the basis of soil type, and length and steepness of slope on terrain categories) and a digitised vegetation map (cover values based on 1988 estimates of natural vegetation and cropping areas). Yet the map is predominantly dark red (for danger!) and it seems likely that most readers would assume a very serious situation. Certainly the conclusions that erosion is occurring more rapidly than soil formation, and that it has accelerated since European occupancy, are correct. Yet this published map includes data showing natural erosion rates in the tropics of 500 t/ha/yr, a figure comparable to the extreme erosion rates on sugar cane farms in the wet tropics (Johnson et al. 1998). Unless we heed the caveats in the caption, this and similar maps give a false impression.

Furthermore, the extent of degradation in any area is not uniform. For example, R. F. Isbell (1986) cites several studies from non-arid parts of Queensland and the Northern Territory that estimate that 11–12% of catchments are affected by severe erosion. But nearly all the affected areas are on particular soils—that is, the solodic soils that have highly dispersive, and thus highly erodible, subsoils. About half the affected areas are located where these soils are shallow and the slope gradient is moderate. Generalised assessments may mean little if they obscure these details. Certainly management and remediation strategies need to be based on an understanding of the processes and particular characteristics of affected areas, rather than just a broad percentage viewpoint.

A forceful monetary analogy illustrates the severity of soil erosion problems (Figure 3.3). It may seem obvious that eroded topsoil would be replaced by soil forming at lower levels. As long as the rate of erosion is matched by the rate of soil formation, then there is no real loss. Until recently quite high rates of soil formation were thought to apply: rates of 2.5 cm per year if the soil was ideally managed, and of 2.5 cm per 30–100 years under natural conditions. Judging by these figures, agriculture and soil management seemed to be promoting good soil development. Unfortunately, even if these rates are valid for the young, silt-rich soils of North America, for which they were measured, they do not apply to the old, weathered soils of Australia. Here, where the topsoil (A horizon) is stripped off, there is no evidence that the subsoil (B horizon) is turning into productive soil to replace it. The surface of the soil may be darkened by organic matter, but the depth of soil does not increase as fast as erosion removes soil. We are squandering our 'withdrawals' from a diminishing 'account' (Beckman and Coventry 1987).

Our 'spending' can be increased by natural fluctuations, such as floods and droughts. A study of erosion on the floodplain of the Snowy River, near Lake Curlip in Victoria (Boon and Dodson 1992), showed that erosion before European settlement in the early nineteenth century was associated with charcoal deposition, probably from Aboriginal burning operations. Since then erosion rates have increased. Vegetation changes have meant perhaps more intense fires, and clearing has exposed soil to erosion. The most severe erosion followed an intense fire in 1890, and subsequent heavy rainfalls and flooding from 1890 to 1893.

In other situations it has been drought that has worsened the damage. Droughts have a major impact on agriculture and land degradation in Australia. They are not easy to define, because so much of the country is arid and can thus expect low rainfall at most times. Crops and introduced pastures are not adapted to seasonal aridity, so the effects of low rainfall can be more severe in the major agricultural zones of southern Australia than in the pastoral zones dependent on native pastures. For statistical purposes, an area is considered to

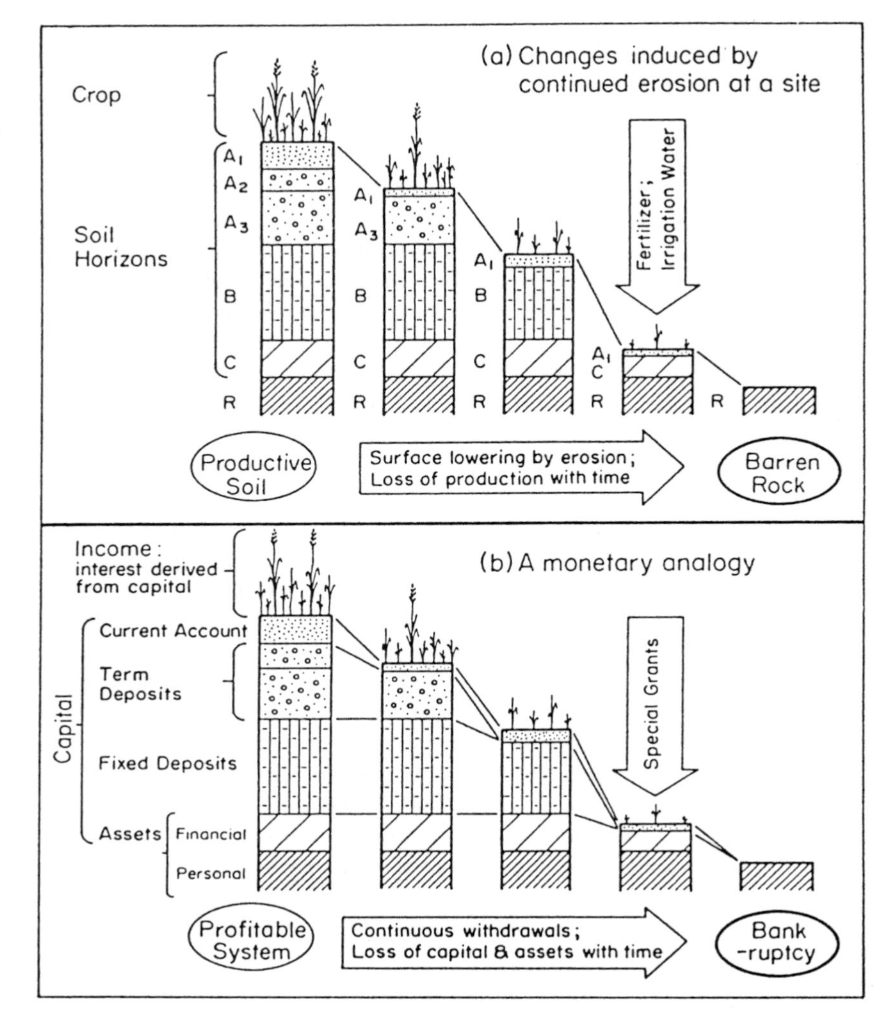

Figure 3.3 A monetary analogy illustrating the loss of valuable soil horizons, and the difficulty of replacing eroded soil. (Source: Beckmann and Coventry 1987.)

have deficient rainfall when, for three months or more, it receives rainfalls lower than the lowest 10% (first decile) of historically recorded falls for those months. Major droughts have been common since settlement (Table 3.1), and their impacts can be illustrated by the effects of the 1982–83 drought in southern Australia. Total losses were estimated in excess of A$3000 million. Widespread bushfires in Victoria and South Australia culminated in the terrible Ash Wednesday conflagration of 16 February 1983. In the same month exposure of soil to wind erosion in the Mallee led to a huge dust storm, which engulfed Melbourne and dropped nearly 10 kg of soil per suburban block within an hour (Beale and Fray 1990, McTainsh and Boughton 1993).

Table 3.1
Major droughts in Australia

Drought period	*Extent*
1864–66	severe in eastern Australia, from Queensland to Victoria, and in Western Australia
1880–86	affected the eastern part of central Australia, from Queensland to South Australia
1888	affected all States except the Northern Territory
1895–1903	probably Australia's worst drought; sheep numbers were halved; the whole country was affected
1911–16	the whole country affected for some of these years
1918–20	affected the whole country
1939–45	the whole country affected for some of these years
1958–68	widespread and probably second only to the 1895–1903 drought in severity
1972–73	affected eastern Australia
1982–83	affected nearly all eastern Australia, especially the south-east; record low rainfalls for 11 months
1991–92	affected mainly Queensland but reduced agricultural production in Australia by 10%
1997–98	widespread in Australia and also Indonesia

Data from ABS 1982–99, and Bureau of Meteorology
<http://www.bom.gov.au/climate/drought/>

Wind erosion

Not surprisingly, the most severe wind erosion occurs during droughts. At the end of the 1895–1903 drought a huge series of dust storms engulfed Victoria and parts of New South Wales, Queensland and South Australia over a three day period from 11 to 13 November 1903. Many places experienced gales of dust, fire balls, lightning, and darkness during the day that was so intense that the fowls roosted (Noble 1904).

The drought of 1939–45 led to severe wind erosion in South Australia, New South Wales and Victoria, but—as mentioned in Chapter 2—increased windiness also contributed. In the 1958–68 drought the number of dust storm days was lower than in the earlier drought, and since early 1970, the number of dust storm days has been very low. Wind erosion seems to be less of a problem now than in earlier times, but the reasons are not entirely understood. Natural fluctuations are certainly significant; recovery of land due to better rabbit control and changed farming practices may be a factor; but part of the explanation may lie in the worsening of another form of land degradation: the spread of woody weeds (McTainsh and Leys 1993). These have occupied many overgrazed areas in the semi-arid parts of the country, creating a 'green desert' that hinders wind erosion but drastically reduces productivity, a process criticised in forthright terms even in the 1940s:

> Such weeds as 'Lincoln weed' and 'Jockey-bush', or 'Paddy's Lucerne', have recently been hailed as potential saviours of the eroded country. Arrant, unscientific rubbish. True, they may hold the drift for a while, but if the present rate of grazing is maintained, they too will disappear, to be replaced by still more inferior types of flora.
>
> Pick 1942, p. 88

However, wind erosion is not a problem that has disappeared. In late May 1994 a huge dust storm moved across the southern half of the continent (Bureau of Meteorology 1994). An intense low pressure system caused storms that battered the Perth area with wind exceeding 56 km/hour for 17 hours on 23 and 24 May. Strong winds ahead of the front created dust storms that affected all districts of South Australia on 24 May. Airports were closed and an estimated 20 million tonnes of soil was moved. Dust spread across New South Wales, and into far south-west Queensland and northern and central Victoria, on 25 May. Visibility was reduced to less than 100 m in some places, and dust was still widespread in the eastern States on 26 May. A few days later, on 30 May, strong northerly winds ahead of another front caused almost as severe dust movement again, in South Australia (Figure 3.4).

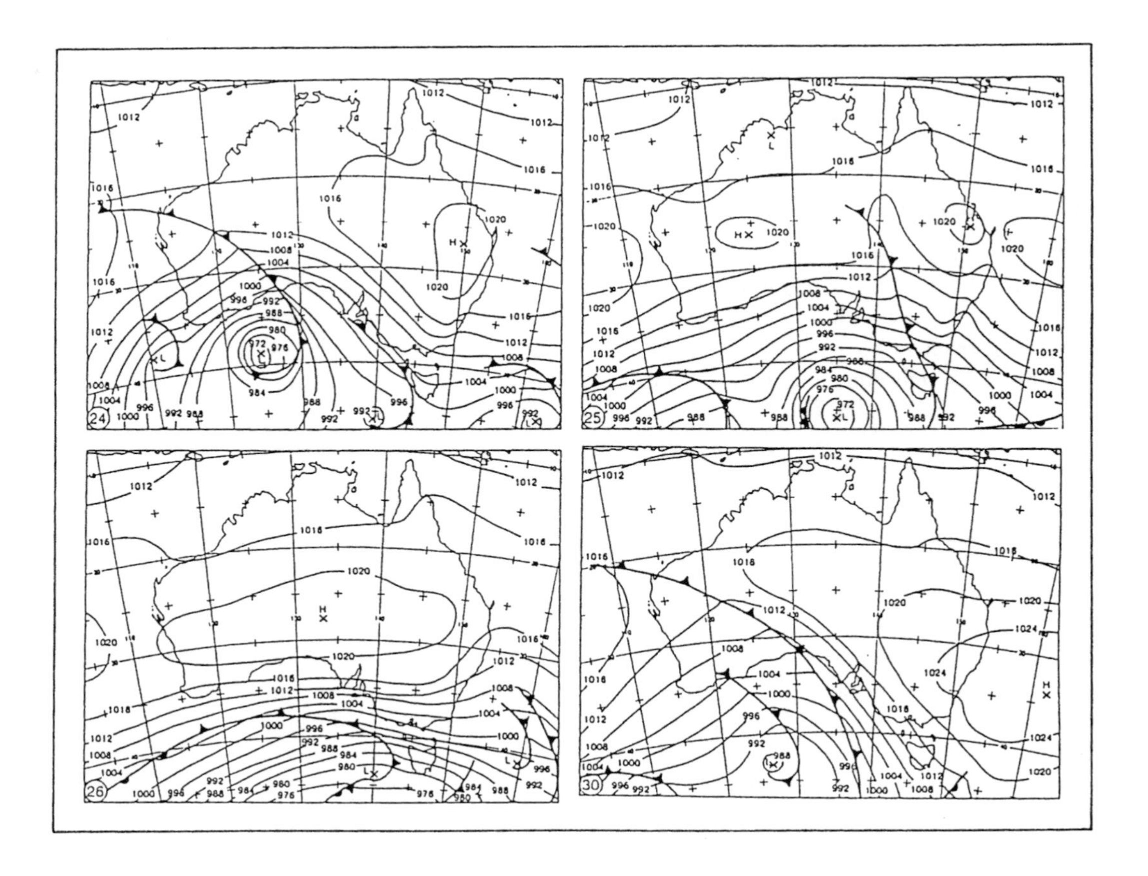

Figure 3.4
Pressure systems moving across Australia from 24 to 26 May and on 30 May 1994. The intense low pressures and strong cold fronts caused strong winds and severe dust storms. (Source: Bureau of Meteorology 1994.)

Wind erosion depends on wind velocity, the erodibility of source material, the soil moisture and the protective cover of vegetation. Wind selectively winnows out organic matter and fine particles. Particles about 0.8 mm in diameter require the least energy to be picked up; coarse sands are moved only over short distances, and clays may be too cohesive to erode easily. G. H. McTainsh et al. (1990) calculated an index of potential wind erosion (E_w), based on climatic factors. These factors were mean annual wind run (an estimate of velocity) and Thornthwaite's index of precipitation minus evaporation (an estimate of soil moisture). Then E_w was correlated with wind erosion (estimated by dust storm frequency between 1960 and 1984) for eastern Australia. Throughout Australia wind erosion is most prevalent in the drier regions, and most dust storms occur outside the main agricultural and pastoral lands. However, some places in the eastern-Australian study showed erosion rates above the level explained by climatic variation (Figure 3.5). This accelerated erosion was presumed to be due to the influence of human activity.

Four areas of accelerated erosion were identified. The Mallee–Riverina region of New South Wales has a long history of wind erosion, both before and after settlement. Its solonised brown soils are highly erodible, and clearing and cultivation has exposed them to severe erosion. The Charleville region of Queensland has sandy red earth soils, and clearing of the mulga is exposing these to wind erosion. The other two regions—around Longreach and around Camooweal in Queensland—have clayey soils, which are not usually easily eroded by wind. However, in these areas the clays often pelletise to form sand-sized aggregates that wind can entrain. In these regions occasional storms, rather than land-use change, seem to be the main cause of severe wind erosion (McTainsh et al. 1990). Nevertheless, land-use change has made significant impacts on vegetation cover, which have had resultant effects on erosion.

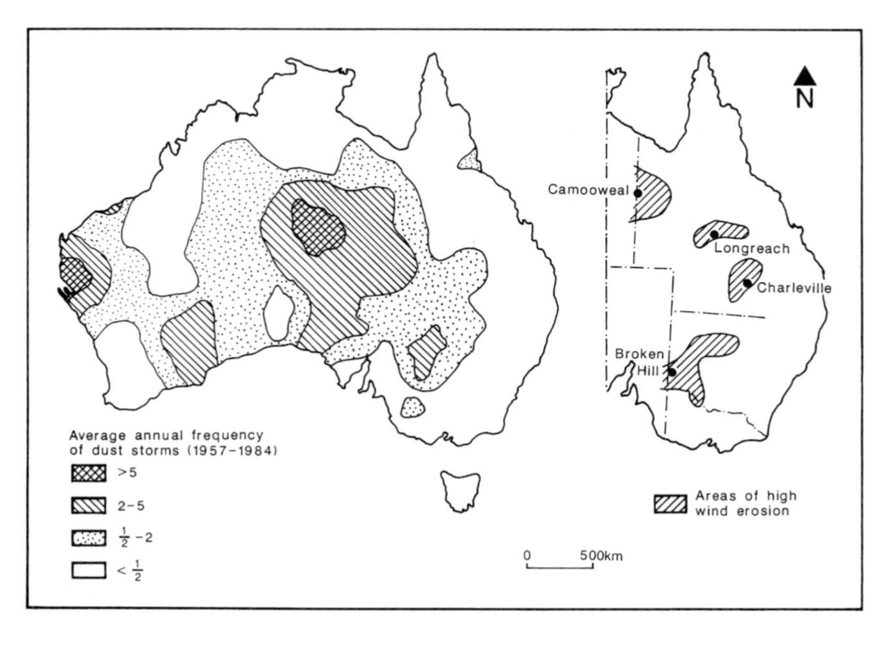

Figure 3.5
Average annual frequency of dust storms, and areas where wind erosion is higher than can be explained by climatic variables. (Source: redrawn from McTainsh and Leys 1993, and McTainsh et al. 1990.)

The impacts of pastoralism and cropping

As sheep and cattle spread over the country, there were drastic changes in the vegetation cover. Trees were cleared by axe and fire. As softer and more palatable grasses were grazed selectively, coarser grasses like spear grass became dominant. Pastoralists soon learnt that burning these forced the release of nutrients and the growth of 'green pick', on which their animals thrived. Annual burning became a standard tool of management, and remains so, especially in northern Australia. Around small holdings, fires became less common than they had been under Aboriginal or squatter occupancy. Fires were a nuisance in fenced runs. But excluding fires allowed woody shrubs to invade. Dense thickets took over land that had been grassy and sparsely tree-covered (Pyne 1991). Erosion as a result of overgrazing increased the extent of bare areas, so that less rainfall seeped into the soil and grasses did not grow well in the drier soil conditions. A higher percentage of infiltrating water entered the soil via insect holes or dehydration cracks, to depths below grass roots but available to woody shrubs. In the mulga country of south-west Queensland woody shrubs that stock will not eat (for example, *Eremophila* spp.), or thickets of mulga with no grass understorey, are still spreading into pastures that were once grassy and lightly timbered. Seventy properties in the area bounded by Charleville, Quilpie, Thargomindah and Cunnamulla were surveyed in 1985. Of these, 54 had over 60% bare ground, and 51 had more than 3% covered by woody weeds. In that region a woody weed cover of 3–6% reduces pasture productivity by 20–30%. None of the 70 properties had better than mediocre land condition (Cameron and Blick 1991).

Grazing animals also have impacts on the land and the conventional wisdom is that more damage is caused by the harder hooves of cattle and sheep than by the softer-footed native marsupials. J. C. Noble and D. J. Tongway (1986b) dispute this. They suggest that both types of animal exert about the same pressure on the soil, but that damage occurs because the number of domesticated animals is so much larger. Particularly after fencing wire and artesian water supplies became commonly used from the 1880s onwards, stock concentrated around watering points, and paddocks carried more stock than they could support in the long term. Erosion began and was most severe in places where stock congregated, near water sources, yards and gates. The plant cover was eaten out; the soil was compacted and pulverised; the patches where urine and faeces were excreted were unsuitable for plant growth. Nutrients were lost from the whole paddock by export in meat, wool and hides, and were blown away in the smoke from burning.

As topsoil was washed away, more nutrients were lost. Where agriculture and pastoralism rely on shallow-rooted species, the nutrient content of the topsoil is crucial to productivity. We have become familiar with the idea that soils under rainforest are not necessarily fertile and that recycling of nutrients from leaf litter is vital to the health of the rainforest. However, many people are not aware that the same process is vital under less glamorous vegetation communities, and on a wide range of Australian soils. Under saltbush and mallee in western New South Wales, nutrients are rapidly recycled as leaves are shed and then decomposed. If the top 10 cm of soil are stripped off by erosion, more than one-third of the soil's store of nitrogen and organic matter is lost (Burch 1986). More erosion can expose a saline or sodic B horizon, leaving a bare saline scald with, at best, a fringe of halophytic succulent plants.

After clearing, soil organic matter and thence organic carbon contents usually decline. Less litter is added, microbial decay of organic matter is faster, organic-rich topsoil is mixed deeper into the soil. In northern New South Wales, carbon contents of cropped soils were less than 50% of contents on uncropped reference sites, and much of the decline came within the first few years of cropping. The hydraulic conductivity (soil drainage) and aggregate stability (soil aeration and structural stability) fell with the declining organic contents (Whitbread et al. 1998).

During rainfall more water runs off bare ground than off vegetated ground, and sheetwash removes more soil. This can be estimated using the Universal Soil Loss Equation (USLE), a technique that has been modified from its original development in the USA and adapted to Australian conditions (Charman and Murphy 1991, Rosewell 1997). The equation is an empirical relationship, written as:

A = R*K*L*S*P*C

where

A = average soil loss (tonnes/ha/year)

R = rainfall erosivity, a measure of the power exerted on the ground by rain, and estimated from rainfall intensity and duration

K = soil erodibility, which depends on the texture and structure and organic content of the soil

L, S = slope length and gradient

P = effect of soil conservation measures in cropped lands, due to factors such as the spacing and depth of contour banks

C = effect of the type and density of vegetation cover.

The USLE has one major flaw for Australian conditions: it does not estimate the loss due to subsurface erosion or gullying. Clearing, overstocking and other factors that reduce the vegetative cover accentuate these forms of erosion, and we have seen already their significance in Chapter 2 (Box 1). Gullies extend headward as overland flow removes material, or by collapse after the subsoil is eroded by splash or by piping. They widen as overland flow washes into them, as raindrop erosion shifts material into the channel, and as throughflow or channel flow undercuts the edges (Charman and Murphy 1991). In both pastoral and cropping areas gullies may be the most significant form of land degradation. We need to consider not only the removal of vegetation by grazing by domestic stock but also the influence of introduced pests and native fauna.

Rabbits and 'roos

Rabbits spread rapidly across Australia after wild greys were introduced for sport to Victoria in 1859. By 1880 they had caused such havoc that they were declared illegal, and feral cats were protected to prey on them. Rabbits ate out pasture, nibbled off tree suckers and seedlings, initiated tunnel erosion, sheltered in blackberry infestations and encouraged weeds like St John's wort by restricting pasture regrowth. Farmers herded them into corners of fences and clubbed them to death by the hundreds. Poisoning and trapping were to no avail, and they spread throughout southern Australia (Barr and Cary 1992). Sheep were outcompeted, and some farmers had to abandon their runs. Heat or starvation killed the rabbits by their thousands in the inland, and corpses piled up in putrid masses. But still the pests advanced. By 1900 they had spread out of Victoria, across New South Wales and South Australia; by 1910 they occupied Queensland, the Northern Territory and coastal Western Australia as far north as the tropics (Lines 1991).

They were less of a problem in cropping areas, where cultivation disturbed their warrens, than in grazing country. The cost of controls such as trapping were too great in the extensive properties, such as those along the Darling River, and even landholders who spent large sums found they had more rabbits at the end than they had at the start. Yet not everyone was united in the attack. Overlanders who used them as a source of meat continued to spread rabbits intentionally. By the 1940s trapping for meat and fur was conducted on such a large scale that the rabbit industry strongly resisted attempts to bring in rabbit disease control, even though the value of the industry was only one-tenth of the estimated loss caused to pastoralism:

> The rabbit industry is . . . very militant in guarding its own interests . . . The sentimentalists add their weight . . . In the meanwhile the rabbit goes unmolested on his own sweet path of destruction . . . It would take an eye prejudiced by either ignorance or self-interest to discern in the trade in furs and carcasses any real compensation for this stupendous loss.
>
> Pick 1942, pp. 40, 41

Today, rabbits number 200 million and eat enough feed for four million head of cattle (Beale and Fray 1990). Yet they receive only brief mentions in government surveys (Graham 1989, Woods 1984) and books (Beale and Fray 1990, McTainsh and Boughton 1993) dealing with land degradation, and in exhortations for sustainable rural development (Cameron and Elix 1991).

Their numbers plummeted in the early 1950s after the release of 1080 poison and, more particularly, the myxomatosis virus. In southern Australia 95% died and there was a dramatic increase in the 1952–53 wool clip as a result. But in the inland, the virus did not spread as well, and many rabbits survived. In 1990–91 they erupted in South Australia and ate the ground bare in a strip several km wide for 200 km along the Dingo Fence. They then died by the thousands from disease, starvation and heat stress. Unfortunately in the wetter areas there was little follow-up to the initial success with myxomatosis; the rabbits developed resistance to the disease, and their numbers rapidly built up again. A new vector, the European rabbit flea, was introduced in 1969 and again numbers fell. Where there was follow-up action, the problem was kept under control. For example, on one property rabbit numbers fell in 1969, but recovered to more than their previous level by 1990 when no action was taken. On the adjacent property, where the warrens were ripped, rabbit numbers were lower initially and continued to decline after the flea was introduced (Cooke 1993). The latest strategy-release of calicivirus is having variable success.

Controlling rabbits is important not only to increase productivity and decrease erosion; rabbits also play a significant role in reducing tree growth and in displacing native fauna. Simply excluding livestock from grazed areas does not allow regeneration of woodland, such as the Victorian Mallee and the *Acacia* woodlands of South Australia, unless rabbits are controlled. Introduced rabbits in the inland competed for food during droughts and hastened the extinction of native mammals. The distribution of the bilby has been reduced to 20% of its original range, and it now survives mainly north of the Tropic, outside the area occupied by rabbits. Hence, the Foundation for a Rabbit-Free Australia (formerly the Anti-Rabbit Research Foundation) has adopted the slogan 'Bilbies not bunnies' to try to counteract the vermin's appealing and cuddly image (Figure 3.6). As with many environmental problems, the solution is complex. Rabbits are the main prey of another introduced pest: the fox. If rabbit numbers are controlled, foxes may turn to native animals for food (Cooke 1993). The ways forward must include a multifaceted attack on rabbits, but also measures to control foxes.

Foxes and feral cats have killed many small native mammals, and the rabbits have occupied the burrows of these mammals. In north-western New South Wales CSIRO researchers have found evidence that woody weeds have spread partly because of fewer fires, but also because these introduced animals have displaced the bettongs, or rat kangaroos, that used to feed on the seedlings of the weeds (*The Land* newspaper, 28 July 1994).

Figure 3.6 Cartoons and logo used by the Anti-Rabbit Research Foundation (now the Foundation for a Rabbit-Free Australia Inc.).

Although rabbits cause the most obvious damage, most other introduced animals have also become feral and now cause significant problems, particularly in pastoral areas. Cattle, horses, donkeys, camels, buffalo, pigs and goats compete with grazing stock, and foxes and feral cats prey upon native fauna as well as rabbits. However, it is not only introduced fauna that compete with stock. Competition from kangaroos and other macropods is considered by many graziers to be a major problem. Certainly, experiments around Charleville in Queensland showed that excluding sheep only did not allow recovery of native pasture, but excluding sheep and kangaroos led to good recovery. There is direct competition, particularly during drought, and in marginal sheep grazing country and on lands stressed by overgrazing. Kangaroo populations have increased since settlement because dingo predation has been controlled and because permanent water supplies have been provided for livestock. Kangaroos have taken advantage of these changes. Current control practices are based on shooting, and most kangaroos shot are adult males. This does not lead to population control. Wild harvesting of kangaroos has been proposed, as has a more comprehensive government-controlled shooting program. It is a controversial issue: some people believe it is immoral to shoot the kangaroos; others see control as necessary; still others see the kangaroo as a harvestable and renewable resource, preferable to sheep in marginal rangelands. However, the real need, from the point of view of total land management—as J. Cameron (1991) emphasises—is to quantify and balance:

- the total grazing pressure (livestock, feral animals and native fauna)
- the sustainability of soil and water resources
- and the conservation status of native species and communities.

Irrigation and salinisation

While dryland agriculture is the most extensive use of land, the demand for irrigation continues. The economic reason for this is clear: although irrigation occupies only 0.5% of agricultural land, 21% of estimated value from agricultural operations in 1989–90 was derived from farms with some irrigated area. High productivity comes at a cost, however, as not all water added to the landscape by irrigation is used by plants. Some seeps down into the soil and can cause the water table to rise. In semi-arid areas the water table may be saline because of concentration of salts by evaporation over geological time. Seepage to groundwater from irrigation can add to this saline water table. Salty water rises closer to the surface and even up to the root zones of plants. The water table may still be a few metres below the ground, but saline water may be drawn up into the root zone by capillarity (that is, the rise of water through narrow pore spaces, due to the force of surface tension, in the same way that water rises up a very fine tube). Evaporation concentrates salts in the topsoil, and plants cannot survive. With more irrigation and further rise in the water table, saline water comes right to the surface, and saline seepages begin. Both the soil and the streams are affected by this salinity. Also, irrigation adds salt directly to the soil because water used for irrigation has some salt dissolved in it. The plants do not take up this salt and it remains in the soil, or is flushed to streams or groundwater.

The effects are greatest where water tables are naturally close to the ground surface and in areas where the soils are already saline. In the Murray–Darling Basin, the dominant soils used for irrigation are red-brown earths and grey, red and brown clays. Even before irrigation began, these soils often had high levels of salinity and sodicity in their subsoils (Box 3). The crops grown also influence the impact of irrigation. Between 1950 and 1980 the area sown with rice in the Murray and Murrumbidgee irrigation lands rose from 15 000 to 110 000 ha. Rice crops are flooded for four months a year, and about 25% of the water put onto the paddy fields seeps down to the water table (Watson 1986). Thus, in some districts, rice has been restricted to heavy clay soils with low permeability, and the area sown and the number of crops has been limited.

How does salinisation affect soils and plant growth?

- Too much salt in the soil or soil water can cause reverse osmosis, and dehydrate plants. Sometimes the saline water contains toxic levels of some micronutrients.
- As clays become sodium-saturated, they no longer flocculate. The very fine clay particles no longer clump together, but act as individual small particles, and are easily transported by water flowing over or through the soil. Soils become dispersive and easily erodible when wet, and crusted and flaky when dry. When wet, they form an impervious surface layer, so water does not penetrate easily. The root zone may receive little fresh rainwater. Most rainfall runs off quickly, causing erosion.
- Dispersive clays, and the compaction of wet soil by machinery during cultivation, lead to poor soil structure, so both water and air move through it less freely.
- Waterlogging of the soil reduces the aeration of the soil. This discourages biological activity and soil fauna, and consequently can reduce soil fertility and degrade soil structure.

The Murray–Darling Basin is the source of 45% of the gross value of agricultural production in Australia. Its water resources are more developed than any other drainage basin in the nation, and it supplies about 60% of water consumption. Its importance to irrigation is even greater. Irrigation uses 90% of net water consumption in the Basin, and over 70% of Australia's irrigated hectares lie within the Basin (Figure 3.7). The Basin is also one of the nation's most difficult natural resource problems (Crabb 1988). The problems arise because of the water balance in the Basin. Irrigation depends on fresh water collected in the high-rainfall margins to the east and south. This is stored behind dams and weirs, and released down canals and the major streams. The relatively fresh water delivered down-valley supports not only irrigated agriculture, but also an important tourism industry.

The high-rainfall areas on the eastern edge of the Basin are also the zones where water is added—by infiltration deep into the strata and sediments—to recharge the groundwater resources of the Basin. On the areas of low elevation in the western and central part of the Basin, this groundwater may intersect the ground surface and discharge into the streams. The Murray–Darling Basin is a closed groundwater and sedimentary basin. The sediments are thickest in the central western part of the Basin, so groundwater entering along the recharge

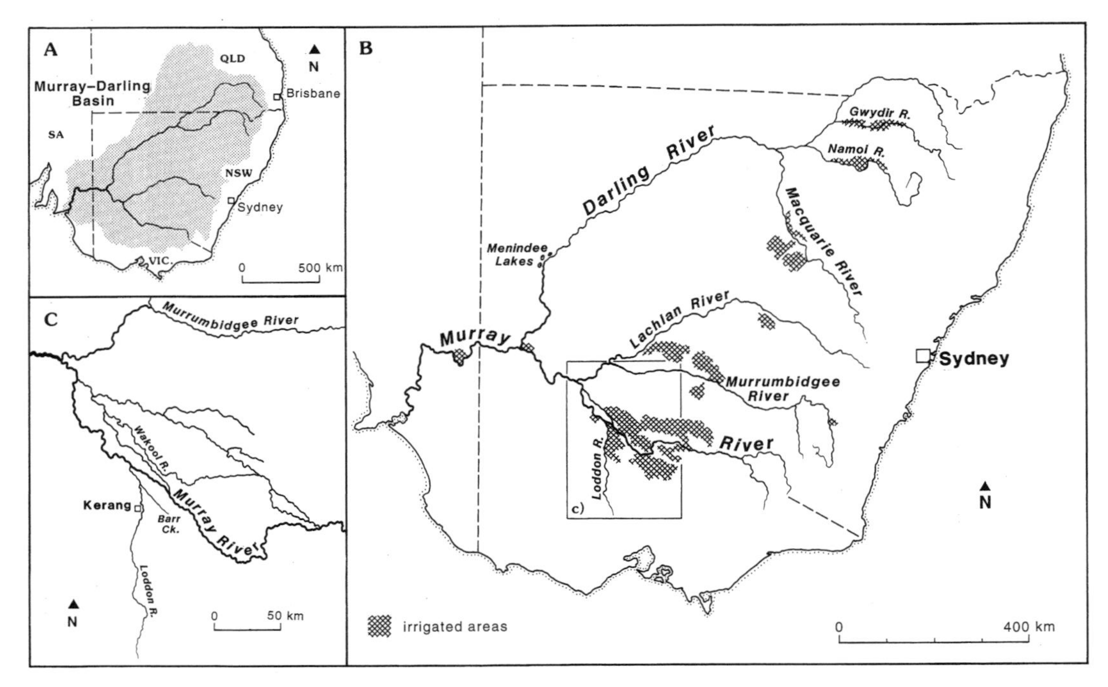

Figure 3.7
The Murray–Darling Basin and its irrigation areas, with detail of the Loddon region. (Source: redrawn from Simmons et al. 1991, and Mackay and Eastburn 1990.)

zones of the eastern highland rim is trapped within them. The sediments are therefore largely saturated. Groundwater is saline (except near the recharge zones and in the limestone aquifers of the south-western section), partly because of leaching from Pliocene marine deposits, but mainly because of evaporation. Water from the Basin is lost only by evaporation (where the water table is close to the surface) and by seepage into streams. There is so little capacity for extra groundwater to be stored that any increase in infiltration prompts a rapid rise of the water table (Evans et al. 1990). Yet irrigation imports 200–600 mm into parts of the Basin where average annual rainfall is only 300–500 mm.

In the Loddon and Wakool valleys, irrigation has been developed where groundwater is naturally close to the surface and discharges into the rivers (Macumber 1990). When Major Mitchell first crossed the Tragowel Plains, near Kerang on the Loddon River, in the winter of 1836, they were treeless and covered in lush grass. In a historic misjudgment he wrote that they were fertile plains, which could be improved by canals to better spread the water across them. Others coming soon after found them bare during dry seasons and noted the extent of salt-loving pigface, but this did not deter settlement. Building of levees ponded water on the fields in winter, and the water table began to rise. By the 1920s the water table had risen from between 8 and 10 m below the ground surface to within 2–3 m, and salt water was discharging along some channels. This did not quell enthusiasm for irrigation, and more water was seen as the key to making a marginal area profitable (Barr and Cary 1992). Salinity in the area is now severe, with a 1991 study showing 48% of the irrigated land in the Loddon–Avoca region to be salt-affected (Eberbach 1998). However, careful management can reduce soil salinity to acceptable levels even in this difficult environment. At the Kerang Agricultural Research Farm the land was smoothed and drainage channels reconstructed to shed water rapidly from the fields. A dewatering pump was installed in 1964. Plants were used to reduce soil moisture; cultivation was done in autumn when water tables were lowest; and some additional irrigation was used to flush salt through the soil. The percentage of salt (as NaCl) in the top 60 cm of the soil at over 100 plots on the Farm dropped from between 0.5 and 0.7% in 1961 to 0.25–0.4% by 1967; it then stabilised at 0.15–0.25% from 1973 to 1988 (Jones 1990).

Nevertheless, salinity reduces the productivity of about one-third of the Kerang Irrigation District (Watson 1986), and Barr Creek, which drains the District, is the largest point source of salt entering the Murray River. Even though an interception scheme was built in 1968, only 20% of the salt from the district is diverted; the rest flows into the Loddon River and then into the Murray, causing a sharp increase in salinity (Figure 3.8) (Close 1990). What has happened in the area illustrates well the delicacy of the water balance in the Murray–Darling Basin. The years 1973–75 were unusually wet, and the groundwater pressures in the sediments jumped between 3 and 4 m in the Loddon Valley. Saline seepages spread out of the irrigation area and 20 km upstream as the extra rainfall infiltrated and forced groundwater to discharge over a wider area. Neither extensive pumping-out of groundwater nor the severe droughts of 1982–83 have reversed the damage (Macumber 1990). Unfortunately salinisation is not only a significant issue in irrigated areas; dryland salinisation is also a major problem.

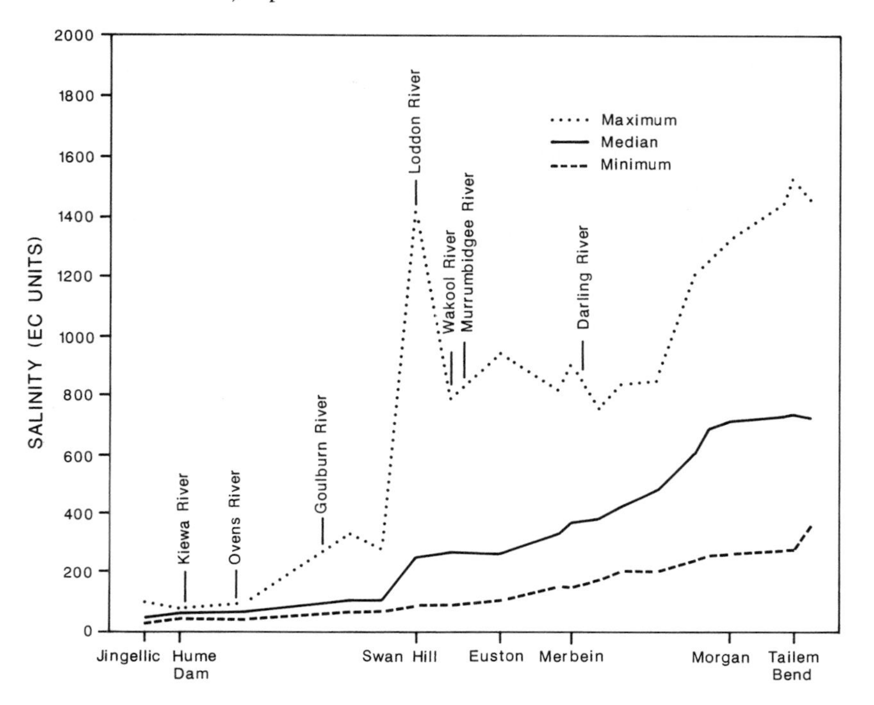

Figure 3.8
Line graph showing the increasing salinity in the Murray River, moving downstream from Jingellic to Tailem Bend in South Australia. (Source: redrawn from Mackay and Eastburn 1990.)

Box 3
Soil salinity and sodicity

Salt-affected land has been defined as land that is saline, or sodic or strongly alkaline, anywhere within the top metre of the soil profile. What do these terms mean?

- Saline soils have accumulations of salts, mainly sodium chloride, because evaporation has been higher than rainfall or because salty water has been evaporated from them. They are natural features over much of arid Australia, and salt-tolerant succulent plants may occupy the edges of bare clayey salt pans. The level of salt needed to affect plant growth and to characterise saline soils is not very high: levels in the upper horizon of >0.1% sodium chloride in coarse-textured soils, and >0.2% in fine-textured soils, or a level in the subsoil of >0.3%. The more saline the soil, the higher its electrical conductivity (EC). This is measured in microSiemens per metre (mS/m) on a saturation extract from the soil (EC_e), or on a suspension of 1 part soil to 5 parts water (EC). Soil with 0.1% sodium chloride will have an EC_e of approximately 400 mS/m; the EC value will depend on the texture of the soil, but will be 5–17 times lower.
- Sodic soils have many of the cation exchange sites on their clays saturated with sodium ions rather than nutrients like calcium, magnesium or potassium. This means they are not fertile. It also means that the clays do not flocculate, but are dispersive. When sodic soils become wet, the clays do not stick together and remain cohesive. They flow in suspension in the water, and erosion can be very rapid. Soils are defined as sodic if their exchangeable sodium percentage (ESP) is more than 6, and as strongly sodic if the ESP is more than 15. ESP is the percentage of the total cation exchange capacity of the soil (CEC) that is due to sodium, rather than to other cations, such as calcium, magnesium, potassium, and aluminium.
- Strongly alkaline soils are soils with pH values greater than 9.5. These are high in carbonates.

Definitions from Working Party on Dryland Salting in Australia 1982

Dryland salinisation

In semi-arid Australia soils frequently have naturally saline and/or sodic subsoils. In fact, of the 32.4 million ha of salt-affected land in Australia, 90% is naturally saline. These lands include coastal salt marshes, inland salt lakes and areas with saline soils. Only 4.2 million ha are salted as a result of human activity, and much of this (3.8 million ha) is due to the exposure of saline or sodic subsoils by erosion of the topsoil (Table 3.2). Overgrazing is the main cause of erosion. These salt scalds are most extensive in western parts of both Queensland and New South Wales. The remaining salinised areas are affected by seepage salting (usually called dryland salinisation), which is caused by clearing of native vegetation. This is a particularly damaging problem in the wheat belt of Western Australia (Box 4), and it contributes to the salinity problems of the Murray–Darling Basin. In total, dryland salinisation affects twice the area degraded by irrigation salinity (Working Party on Dryland Salting in Australia 1982). As with other environmental problems, diverse estimates of severity are available. For example, the 1996 *Australia: State of the Environment* report (Department of Environment Sport and Territories 1996) quotes two estimates for the extent of salinised land in Western Australia—a 1992 value of 0.56 million ha and a 1994 estimate of 1.6 million ha rising to 2.9 million ha in 2010. CSIRO (1998) suggests 1.8 million ha of a total Australian extent of 2.5 million ha affected Western Australia, on 1996 figures, and that the total potential area in that state was 6.1 million ha.

Table 3.2
The extent of salinised lands in Australia (areas in thousands of ha)

State	*Natural salting*	*Scalding*	*Seepage salting*	*Dryland salinity (1992)*
Queensland	1 863	582	8	10
New South Wales	2 814	920	4	20
Victoria	100	60	90	150
Tasmania	9	nil	5	minor
South Australia	7 058	1 200	55	400
Western Australia	11 351	335	264	1 600
Northern Territory	5 000	680	nil	no data
Murray–Darling Basin				200
Australia	28 196	3 777	426	

Data on natural salting, scalding and seepage salting from Working Party on Dryland Salting in Australia 1982; estimates of dryland salinity from Department of Environment Sport and Territories 1996

Under natural conditions flows of salt and water tend to be in steady state. Despite often-quoted comments, the major source of salt is from the atmosphere. Large areas of Australia have been covered by seas at various times in the geological past, but it is not true that soils are saline because 'much of the Australian landscape is derived from ancient marine sediments' (Eberbach

1998, p. 76). The inputs of salt are from rainfall and, at a much lower rate, from weathering of rocks. These inputs are balanced by outflows in streams and by groundwater discharge. Where evaporation usually exceeds rainfall, salt may also be stored in the weathered rock and soil (the regolith) or added to groundwater. In semi-arid regions salt has accumulated in the landscape over geological time, both in the regolith and in groundwater. Since settlement, clearing of natural vegetation has perturbed the balance. More water now flows through the regolith, and this mobilises the stored salt. More water can infiltrate, and so the groundwater table rises; it may come close to, or reach, the surface (Figure 3.9). Infiltrating water may flow through buried sediments and leach out salt from these. For example, in the Loddon Valley in Victoria, there are buried former channels (deep leads) that contribute only 10% of the outflow of groundwater, but this flow contains 25% of the salt output.

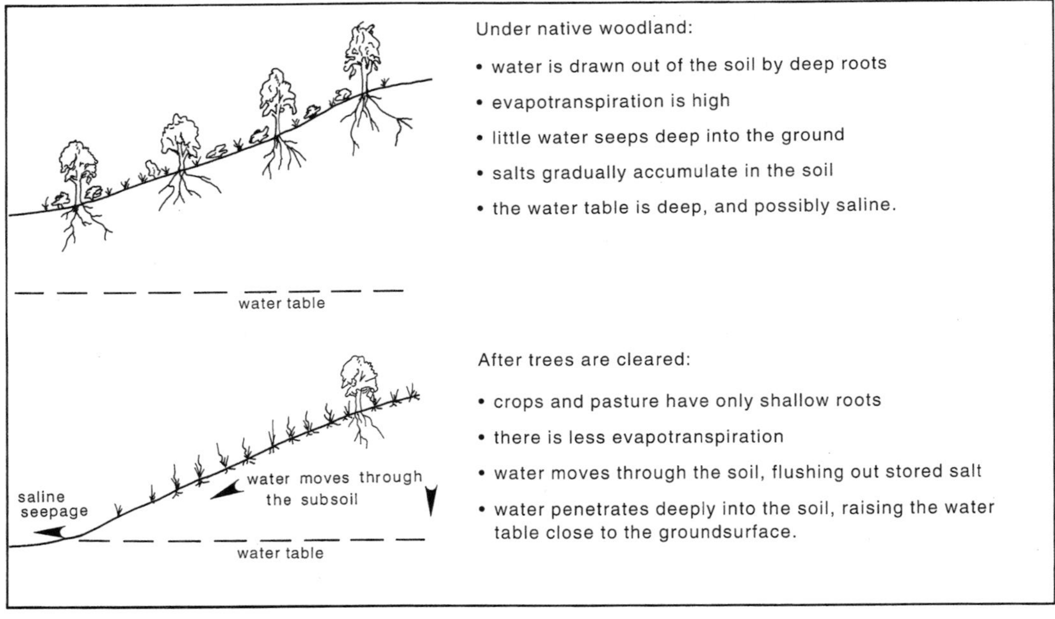

Figure 3.9 Diagram showing the process of dryland salinisation as a result of the clearing of native vegetation.

The native vegetation in northern Victoria and south-western Western Australia was dominated by low, multiple-trunked mallee eucalypts, with a ground cover of deep-rooted perennial grasses. It intercepted or transpired most of the incoming rainfall, so that less than 0.1 mm/year from an annual rainfall of 250–300 mm/year was added to the water table under *Eucalyptus* mallee woodland. The water was extracted from depth, as well as from the surface soil, because roots reached down 10–20 m. This vegetation was cleared and replaced with shallow-rooted and annual pasture and crops, particularly wheat. Evapotranspiration rates dropped, and more water seeped through the soil and drained down to groundwater than previously. After clearing of mallee, recharge rates were rarely below 3 mm/year and sometimes exceeded 50 mm/year, with the mean value being 17 mm/year (Allison et al. 1990). The effect of this change in hydrology was accentuated in Western Australia and Victoria because they have a Mediterranean type of climate, with cool wet winters and hot dry summers. The shallow-rooted annual crops, growing mainly in the warmer months, used far less water than the deep-rooted native perennials. Within a period of between five and 20 years after clearing, seepage salting began. In New South Wales, where rainfall is spread more uniformly, the delay between clearing and the appearance of salt was longer, about 50 years. Salinisation in New South Wales appeared after record rainfalls in 1950, and since then has expanded and contracted with wet and dry seasons (Working Party on Dryland Salting in Australia 1982). Williams et al. (1997) warn that dryland salinisation could occur in wet northern areas also. In the upper Burdekin catchment of north Queensland, strongly seasonal rainfall means that soil water store is filled quickly in summer and excess water can drain below the root zone. If trees are cleared, to provide more land for beef grazing, deep drainage will occur far more frequently than at present. As the soils have considerable salt stores, and the groundwater salinity is 1500–5000 mg/L, a rising saline water table would be likely.

The disruption of the hydrological balance has occurred quickly, but steady state may take a much longer time to restore. For example, in western Victorian uplands some catchments in fairly high rainfall areas (720–990 mm/year) now discharge 25–140 mm/year of groundwater, in comparison to a pre-clearing rate of approximately 10–70 mm/year. To re-establish equilibrium may take from 60 to 300 years. In drier catchments (530–660 mm/year), where groundwater discharge rose to 3–22 mm/year from pre-clearing levels of 1–2 mm/year, the time to equilibrate is much longer, and estimates range from 1200 to 2700 years (Williamson 1983). This is so because much less water flows through the regolith, and the stored salt takes a longer time to be flushed out.

Under forest or woodland, the rate of evapotranspiration is high enough to keep the soil and weathered bedrock unsaturated, except for brief periods after heavy rain. On cleared catchments the soils are saturated during the winter months. Rainfall initiates drainage and the leaching of salt. An obvious solution to the problem of salting is to replant trees over enough of the affected catchment to drastically reduce recharge and water/salt flow. But how much is enough? In some parts of Victoria the ridges are rocky and fractured, with stony shallow soils, allowing easy pathways for water to seep deep into the landscape. On the hillsides and valley sides there are duplex soils with thick, clayey and poorly permeable B horizons. Recharge to groundwater through these soils occurs at a much lower rate than for the ridges. Over 80% of the recharge may enter through about 30% of the catchment (Dyson 1983). If this is so, then reafforestation of the rocky ridges is a good option. Where there are no rocky ridges, and clayey soils are ubiquitous, there are no easily identified recharge areas or obvious areas to replant with trees. Here an agronomic solution, such as planting with deep-rooted perennial grasses such as lucerne over the whole property, may be the best option. In swampy or alluvial areas, native trees such as swamp paperbarks or river red gums, or deep-rooted and salt-tolerant plants such as tall wheatgrass, may be used (CSIRO 1998). Modelling of groundwater changes by AGSO (Australian Geological Survey Organisation) suggests that widespread lucerne planting could drop the water table around the Lachlan

River by 0.7 m, and a corridor of red gums along the river could cause a cone of depression with a maximum 5 m drop and 50 km radius.

In recent years a new technology has enabled easier identification of salinity levels over large areas and better strategies for control (Dixon 1989). An electromagnetic wave is transmitted through the soil to a receiver, and the intensity of the signal received is proportional to the soil's salinity. Rapid surveys of whole properties can be done without the need for drilling and soil analysis. These surveys can be done on the ground or by air-borne devices. This should mean a more accurate assessment of salinisation.

Dryland salinisation has three major environmental impacts:

- increasing the salinity of soils, and so leading to erosion and poor plant growth
- lowering water quality by causing discharge of saline groundwater into streams
- adding to the problems of irrigation salinity.

Box 4
Dryland salinity in Western Australia

Saline seepages in areas with shallow water tables are the most important cause of salinity problems in Western Australia. Where the country is dissected, in the higher-rainfall coastal strip, salinisation is patchy. It is often related to the presence of clay soils or impermeable rock bars, which confine groundwater and create artesian pressure. In the wheat and grazing lands further east, where valleys are broad, very large areas are affected (Figure 3.10). Here there are no large freshwater rivers. The goldfields of Kalgoorlie have depended since 1903 on a pipeline bringing water more than 500 km from the Darling Range. Salinity reduces the quality of the inland's limited divertible freshwater resources. In fact, concern about salty water (expressed by the railways in about 1910) came earlier than the concern of agriculturalists about loss of production. This lowering of water quality then indirectly affects the better-watered

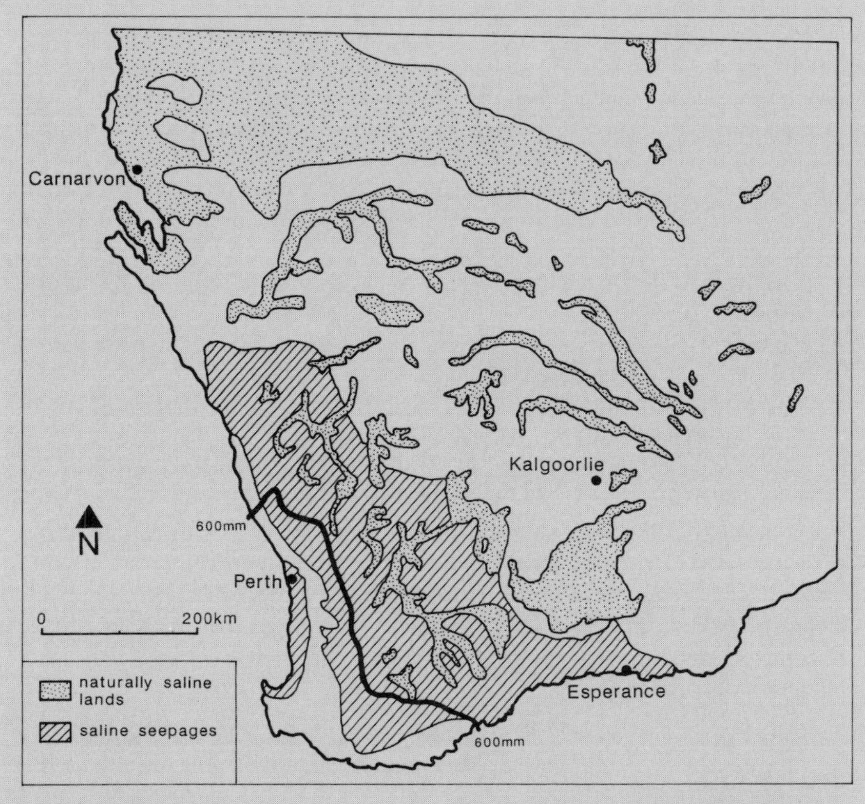

Figure 3.10
Naturally saline areas, and areas affected by saline seepages as a result of clearing, in southern Western Australia. The 600 mm isohyet for average annual rainfall is shown. (Source: redrawn from Working Party on Dryland Salting in Australia 1982.)

coastal fringe, by increasing demand for transfer of water from the major storages in the higher rainfall zones. Salting is a problem that affects large areas both directly and indirectly, and it is a long-term problem. Signs of salting appear about 20 years after clearing, on average, but the stored salts probably take a few decades, minimum, to several hundred years to be flushed from the soils.

Most wheat-belt soils have considerable quantities of soluble salts stored in their profiles. The quantities are highest in fine-textured (clayey) soils that occur on deeply weathered regolith. Groundwater is often saline. The salts have accumulated over geological time from salts blown inland and brought in by rainfall. The saline seepages can develop because of higher infiltration after clearing and the rise of saline groundwater. Water moving down-slope in the soil horizons—a process called throughflow—can redistribute stored salts and flush them to valley floors, where evaporation concentrates them. The valley floors can become waterlogged because of throughflow. By either mechanism—groundwater rise or throughflow—the productive valley floors can become useless wasteland. Seepages also extend up-slope, sometimes almost to the watershed (Plate E).

Many remedial methods have been tried. Simply replanting with eucalypts is not an option in Western Australia because more than half of each catchment may have to be replanted. Deep drainage is rarely feasible; mulching and deep ripping bring little success. Revegetation with salt-tolerant species is helpful, but not all these are useful fodder. Indeed, protection from grazing to encourage water use by plants is an important strategy. Success depends on a combination of methods over a long period. At one grazing property of 4860 ha, work over the last 20 years has included:

- planting 350 000 trees in places protected from stock grazing by electric fences
- planting of tall wheat grass (a perennial grass, which is a high user of water), backed by salt-tolerant trees, along creeks
- construction of 1:60 grade banks that feed runoff into larger drains, to move water quickly off slopes and into creeks
- construction of drains with a W profile, with overburden heaped in the centre, so that runoff is not blocked along their edges.

Salt scalds have been reclaimed and the carrying capacity of the property has been increased by these actions. However, they do not entirely restore affected lands. If fault zones occur in the underlying rocks, water can move more freely along these than through most of the rock. This means that numbers of tree rows or other remedial measures need to be increased to achieve significant reduction of salinisation.

Data from Clarke et al. 1998, Conacher et al. 1983, Lloyd 1992, Malcolm 1983, Working Party on Dryland Salting in Australia 1982

Soil acidification

The acidity or alkalinity of soil—measured by its pH value—is a useful indicator of general fertility. It is correlated with the reserves and availability of important plant nutrients (both macronutrients such as calcium, magnesium and potassium, and micronutrients such as boron and molybdenum); it influences the form and availability of potentially toxic elements (such as aluminium and manganese); it affects the rate and type of weathering of many minerals, including clays, and thus the release of new cations and anions to the soil. In general, most plants, especially those that are agriculturally important, favour a near neutral or mildly alkaline soil. On estuarine deposits, agricultural development can generate particularly severe acidity. This is a special case (Box 5) and the more general problem of soil acidification is widespread in cropping and pastoral areas. Even under natural conditions in Australia, many soils are acidic. Australian soils are old; prolonged leaching has led to depletion of nutrients, and to high levels of silica and aluminium in the soils, particularly in high rainfall zones. A survey of undisturbed soils in parts of Queensland with more than 500 mm average annual rainfall indicated that nearly 80% of the surveyed area had acidic soils (pH <6.6) and 10% of these were strongly acidic (pH <5.6). The same trend occurs in agricultural soils in New South Wales, but the role of parent material also needs to be considered (Ahern et al. 1992, Helyar et al. 1990). In similar rainfall zones soils developed on siliceous materials (such as granites, sandstones or sand dunes) will have lower pH values than soils developed on clays or basalts.

A pH value is a measure of the concentration (or chemical activity) of hydrogen ions (H^+) in a particular substance—in this case, soil. The more hydrogen ions, the more acidic the soil and the lower the value of pH. The theoretical range is 1–14, with 7 being neutral, but in practice, soils very rarely fall outside the range 2–10. The pH scale is logarithmic, so a decrease of 1 pH unit is a tenfold increase in acidity. Comparing pH values from different studies is complicated by the fact that two different methods of measurement are used. The pH values quoted above are pH_w values, measured on a 1:5 suspension of soil:water (Ahern et al. 1992). Other laboratories measure pH_{Ca} values, measured on a 1:5 suspension of soil in 0.1 molar $CaCl_2$ (calcium chloride solution) (Chartres and Geeves 1992). Sometimes the suspension ratio may differ—for instance, a 1:2 ratio is often used rather than 1:5 (Helyar et al. 1990). The values of pH_{Ca} are lower than pH_w values by 0.5–1 pH unit. Thus, direct comparisons of data from different studies may be difficult, but nevertheless, overall trends can still be compared. Certainly the differing measurements do not obscure the fact that many Australian soils are naturally acidic and are becoming more so because of agricultural changes.

Ironically, this has occurred because of a change in farming practice that was seen as the salvation of eroded and depleted soils: the introduction of subterranean clover (sub clover). Adding rock phosphate and superphosphate had contributed to increasing wheat yields since the beginning of the century, but a rapid jump in agricultural production occurred after 1950, when sub clover and nitrogen fertilisers were introduced (McGarity and Storrier 1986). Planting sub clover rapidly increased soil nitrogen levels, which in turn increased pasture growth, improved soil organic matter and reduced erosion. This encouraged rabbits, but the introduction of 1080 poison and the ripping of burrows with new larger tractors reduced this problem. Pasture improvement expanded rapidly in response to high wool prices generated by the Korean War, and after 1963 a bounty for superphosphate further encouraged the trend. 'Clover ley' farming was introduced to the wheat regions. Previously, long fallow periods (which left the soil exposed to erosion) had been used to allow soil organic matter and nitrogen levels to recover after cropping. Now sub clover was planted in rotation with wheat, so there was almost continuous ground cover and lower erosion. Modifications had to be made in low-rainfall areas and to control persistence of root fungi, but varying rotations of pasture, some fallow, wheat and other crops led to high productivity in most areas. In fact, the volume of agricultural production in Australia doubled in a mere 30 years (Barr and Cary 1992). But the desirable increases of nitrogen and organic matter in the soils had the undesirable side-effect of increasing acidity.

Increasing organic matter increases both the amount of complex organic acids in the soil and the cation exchange capacity of the soil. The anionic 'end' of the organic acids are exported from the soil as they are incorporated into plant material that is harvested or grazed. This leaves the hydrogen ions behind to acidify the soil. In the nutrient-poor Australian soils, there may be too few basic cations (calcium, magnesium, potassium, sodium) to fill the additional exchange sites in clays and organic matter, so that more are occupied by hydrogen ions. Hydrogen ions can be added directly to the soil if ammonium (NH_4^+) fertilisers are used, as plant proteins do not use up all the hydrogen from the ammonium ion. However, the main cause of acidification is an indirect result of the increase in soil nitrogen. When bacteria in the root nodules of legumes, such as sub clover, fix nitrogen from soil air, they convert it to ammonium compounds and nitrates. The nitrates that are formed can be leached from the soil, taking with them basic cations and acidifying the upper layers. So the higher the levels of soil nitrogen and organic matter, the greater the yields, and therefore the export, of alkaline products (as grains, meat, and wool); acidity is increased correspondingly by direct addition of hydrogen ions or by leaching of basic cations (Barr and Cary 1992, Chartres and Geeves 1992).

The problem was worsened because the clover and grasses of improved pastures were shallow-rooted. The native perennial grasses were outcompeted, and so rainfall seeped down through the soil. Nitrate leaching led to acidity in the upper layers, and of course, greater infiltration led to salinity problems as well.

What are the consequences of increasing acidity? Low pH usually means low reserves of nutrients (both basic cations and phosphorus). Some crops do not tolerate acidic soils. For example, barley and canola require a pH_{Ca} value above 4.8, and acidification has forced some wheat farmers to turn to more tolerant grains, such as triticale (Barr and Cary 1992, Chartres and Geeves 1992). At low pH values, iron (Fe), manganese (Mn) and aluminium (Al) all become soluble. If the soils are waterlogged, this happens more quickly than if the soils are well aerated. Manganese and aluminium at low concentrations are toxic to many plants. Nutrient cycling by micro-organisms can be impaired in acidic soils. Also, clay minerals will weather, releasing silica and amorphous aluminium oxyhydroxides, which in turn cause hardsetting and crusting of the soil surface. For both chemical and structural reasons, the soil is less favourable to plant growth; yields decline, plant cover is thinner and more erosion is likely.

There are no obvious solutions to the problem. Adding lime is expensive and not effective on all soils, especially those with acidic subsoils. Deep-rooted native perennials, such as wallaby grass, may help, but are not as high-yielding as introduced grasses. It may be necessary to manage properties so that levels of organic matter and nitrogen in the soil are lower, even if this reduces productivity.

Box 5
Acid sulphate soils

On coastal floodplains, development for agriculture can lead to dramatic acidification of soils. This occurs where the floodplains are underlain by old estuarine deposits. In tidal marshes along estuaries, bacteria breaking down organic matter reduce sulphates from sea water, and iron from the sediments, to form iron sulphide, or pyrite (FeS_2). Pyrite is stable while the sediments remain wet and anaerobic. Widespread deposits of Holocene estuarine sediments containing pyrite occur in northern Australia, where they are largely undeveloped, but also along the coast of New South Wales on the floodplains of large rivers such as the Shoalhaven and Tweed. These floodplains have had deep drainage channels cut through them, to speed up the flow of water off them after floods and to lower the water table below the rooting zone of pasture grasses (Plate F). The channels are protected from sea water intrusion by floodgates. In the Shoalhaven area some channels were cut more than a century ago, and were deepened during flood mitigation projects between 1965 and 1972. The lower water table has exposed the pyrite in the sediments to oxidation, and the consequences have been severe.

Oxidation of the pyrite releases soluble iron (Fe^{2+}) and acidity (H^+). The iron may be transported in soil water or in surface flow into drains. It is oxidised to an insoluble (Fe^{3+}) oxyhydroxide, depositing as a red or orange floc and releasing more acidity. This process may deplete the dissolved oxygen levels of the water, and the precipitated floc can smother aquatic plants and organisms living on the stream bed. Very low pH levels result from the oxidation processes. On the Shoalhaven floodplain, the water table close to the surface commonly has a pH of about 3.5. Under such acidic conditions, clays break down, releasing soluble aluminium (Al^{3+}). This is highly toxic to fish and other aquatic organisms. High levels of silica are also released, but the ecological effects of this are unknown. Because the water becomes acidic and deoxygenated, any heavy metals, such as lead or cadmium, that may be present in the sediments are taken into solution.

Not surprisingly, when this highly polluted water is flushed out past floodgates into an estuary, large numbers of fish, crustaceans and worms are killed as a result. Heavy rains cause flooding and turbid water swells the river channels. As the floodwaters recede, and flow-out of the drains predominates, the river suddenly turns bright blue. High aluminium levels flocculate all the sediment, and it drops to the bottom, leaving clear blue water reminiscent of the waters of coral reefs. Thousands of dead fish float to the surface and wash up on the banks. This happened in the Tweed River in 1987 and on Broughton Creek, a tributary of the Shoalhaven River, in 1991 and 1994. The pH of the Tweed dropped overnight from 7 to 4, and the aluminium levels exceeded the toxic level for gilled organisms (0.4 milligrams per litre). A fishing writer described the scene:

> One day the river flowed brown and turbid, loaded with sediment; the next day it was gin clear, with the appearance of a well-maintained chlorinated pool . . . Several days later the carcases floated to the surface . . . Professional fishermen had their nets clogged with hundreds of dead bloodworms which had emerged to die from the river sediment . . . Council trucks removed rotting fish carcases from the river bank where the stench was affecting riverside residents . . . Mosquito larvae . . . were the only signs of life.

Similar fish kills occurred in the Tweed River in 1880, 1916 and 1954, and probably also even before settlement, on occasions when flooding followed prolonged drought.

As well as these dramatic kills, small pulses of acidic water leaking from the floodgates at low tides affect the health of the estuaries. They may force fish away from the gates, causing crowding elsewhere; they probably affect spawning of commercially valuable species, including prawns; they deposit iron-rich floc on the bed; they introduce high levels of heavy metals to the estuary, and possibly the food chain; they seem to be linked to a high incidence of 'red spot disease', which causes ulcers on the skin of fish. Along Broughton Creek, discharge under a 1:1.5 year flood dropped pH levels to below 6 and fish and prawns were not caught for more than a month later.

Thus, the development of the floodplains for agriculture, by lowering the water table and exposing pyrite-rich sediment to oxidation, has led to ecological deterioration in the estuaries.

Data from Bush 1993, Easton 1989, and Pease et al. 1997

Structural decline of soils

In 1989 the Australian government's Standing Committee on the Environment ranked soil structural decline as the most costly form of land degradation in Australia, above erosion or salinisation. It is hard to see how the Committee reached this conclusion, since there has been no national assessment of structural decline and there is little information about its severity or the means of ameliorating its effects. Yet it is a well-known and worldwide phenomenon, particularly in cultivated lands.

Soil structure refers to the continuity, distribution and size of pore spaces between soil particles. Tillage and traffic over the soils, as they are ploughed and as crops are planted and harvested, compacts and shears the soil, altering its structure. The change is greatest in loamy soils. These have a wide range of sizes of soil particles, which can be fitted into the pore spaces as the load from tractors or other machinery passes over the soil. Clay soils are also severely affected because they are often worked when wet, and compaction and shear are increased in wet soils. Decline in soil structure can occur under natural conditions, in hardsetting soils, but agriculture can encourage decline in all soil types (Table 3.3).

Table 3.3
Structural decline of soils

Causes	*Soil factors*	*Load characteristics*	*Processes*	*Effects*
INHERENT PROPERTIES	NO HUMAN INPUT	PRESSURE COMPONENTS	GENERAL	HARDSETTING AND CRUSTING
texture and mineralogy	bare soil surface	surface pressure	aggregate breakdown	massive topsoil
permeability and porosity	intense rainfalls	pressure distribution	slaking; dispersion	surface crusting
organic matter content		period that pressure is exerted	aggregate re-formation	high soil strength
cation status		variability of ground surface	water drop impact	poor root and seed environment
	INDIRECT HUMAN INPUT			
	clearing			
VARIABLE PROPERTIES	organic matter depletion			
water content				
shearing resistance				
bulk density	DIRECT HUMAN INPUT	LOAD COMPONENTS	HUMAN-INDUCED	TILLAGE EFFECTS
	machinery	total load/stress	compaction	massive subsoil ('plough pan')
	domestic stock	vibrational effects	shear	discontinuous pores
	irrigation	rate of loading		reduced porosity

Adapted from McGarry 1993, Table 9.1 and Figure 9.2

Decline may occur rapidly. Plants in a cotton field prepared during dry weather had roots spreading in a dendritic pattern between 0.2 and 0.4 m in depth; on an adjacent field prepared in wet weather, the roots penetrated only 0.14 m and spread horizontally. On clay soils in southern Queensland, cultivation from six months to seven years lowered bulk densities of the soil by 16–28% and lowered organic carbon contents by 28–45% (McGarry 1993). Decline may result from practices forced on farmers by market requirements. In Tasmania potato growers selling on contract to factories may have to harvest even on wet days to maintain supply to the factory, even though they recognise that this is not a good soil management practice (Barr and Cary 1992).

Soil structural decline is a subtle form of degradation, not obvious to the casual observer because it is below ground and difficult to measure. To assess it we need to measure change in soil structural properties, not a simple parameter such as pH for acidity. It is simplistic to assume that more cultivation leads to more decline; the type and timing of cultivation is very important and ploughing under some conditions can improve structure. Structural decline may be blamed for low yields when root disease or infertility is the real cause; or its effects may be attributed wrongly to these other causes. The lack of information is a hindrance to assessing its severity and to designing strategies to ameliorate its effects (McGarry 1993). As with other forms of land degradation, we may recognise the problem, but also be aware of the limitations of our knowledge about it.

Conclusion

In 1942, at the height of spectacular erosion in inland Australia, and in a period of intense patriotism as the Second World War was being fought, there was a call to action:

> At the present time, the best of our young men are on the other side of the world, giving their lives to preserve that heritage of freedom and prosperity which we owe to future generations . . . Of what avail is their sacrifice if an enemy within the gates is allowed to rifle the future resources of their country? Soil erosion is that enemy . . . a destroyer more deadly than all the Hitlers of history . . . We must throw everything we have into this fight.
>
> Pick 1942, p. 89

J. Pick's highly pessimistic predictions have not been fulfilled. The 'erosion decades' of the 1930s and 1940s have not continued. Wind erosion is less severe now than it was then. However, other forms of land degradation need to be addressed: soil acidification and structural decline are recognised; salinisation is widespread; indiscriminate clearing and conversion of pasture to cropland continues. Furthermore, the problems identified by Pick persist. They include:

- poor control of rabbits and the depredations they cause
- inadequate research into, and legislative control of, soil-related problems
- land tenure and financing arrangements that make sustainable land use difficult.

It is not simply a matter of economics. Many landholders take actions that seem economically irrational because of attachment to their land, distrust of new methods, decision-making based on personal experience rather than objective measures, and suspicion of outsiders. While land-use controls are an accepted aspect of urban life, there is strong resistance to controls on rural land use. Rural communities perceive policy changes as being driven by urbanites who do not understand the needs and difficulties of country people. In addition, integrated action and rational decisions are hindered by the plethora of government departments and agencies, at all tiers of government, that have some responsibility for land management. A strongly optimistic note was sounded in the late 1980s as the Australian Conservation Foundation and the National Farmers Federation joined forces to propose the National Land Management Program. The Program was based on the assumption that national assessment of land degradation, consistent assessment of land capability, and encouragement of voluntary, cooperative Landcare groups would enable Australia to achieve sustainable rural land use by the turn of the century (Elix and Cameron 1991). However, the solutions are not simply a matter of good will and education. Poor economic returns in the agricultural sector discourage spending on environmentally desirable practices, both directly and because of related social disruptions. Sustainable use requires action at the regional and ecosystem scale, rather than the farm scale (Pratley and Robertson 1998). The problems are clearly stated by Barr and Carey (1992, pp. 284–5):

> Perhaps our greatest concern is with a widespread belief that the most important task to achieve a more sustainable agriculture is the raising of community awareness and changing of farmers' attitudes to their land . . . What is required are profitable and practical conservation farming techniques and management strategies. Where these are not available the best assistance is research directed at producing . . . solutions, rather than a reliance on evangelical calls to better farming and changing attitudes . . . The dangers of simple prescriptions are that they will not encourage the sustained commitment of the social resources required to continue the unending search for sustainable rural land uses.

Chapter 4

Forestry and Its Impacts

Introduction

AUSTRALIA has not been a well-forested country in recent geological time. The continent drifted northwards towards lower latitudes after the breakup of Gondwanaland, and was subject to major fluctuations of climate far more drastic than any predicted as a result of human-induced alterations (such as the so-called greenhouse effect). Throughout most of the Tertiary period, forests covered much of the continent. In the early Eocene, these were fern–conifer rainforests, but later they were replaced by *Nothofagus* rainforests. This was 'a time of lush greenhouse conditions' (Archer et al. 1998, p. 14). Although wet warm conditions alternated with colder 'icehouse' conditions, the Riversleigh biota in north-western Queensland show that 23–15 million years ago, in the Miocene, this area had a cool wet climate with no dry season that supported lowland rainforests. Rainforest covered areas in central Australia, near Alice Springs, during the Middle to Late Eocene (White 1986). However, by the late Miocene there was a shift to drier conditions which led to the retreat of the continent's extensive rainforest cover and to the march of the sclerophyll vegetation across it. At the start of the Miocene, closed *Nothofagus* rainforest covered the valleys west of the Great Divide in New South Wales, such as the Lachlan. About 15 million years ago, in the mid-Miocene, more frequent fires coincided with a rapid loss of rainforest and replacement by myrtaceous vegetation, probably wet sclerophyll eucalypt forest. Nevertheless, during the Pliocene much of Australia was still wooded, with dry forests and woodlands in the interior and wet sclerophyll forests or rainforests on wetter margins. During the Pleistocene—beginning about one million years ago—the climate became colder, and probably windier and drier. In central-western New South Wales this caused the wet sclerophyll forest to also disappear, and woodland with a grass understorey covered the valleys. The Pleistocene was a time of rapid fluctuations of climate, with at least 20 cycles of glacial/interglacial conditions involving temperature shifts of 5–10°C. The aridity and sparse forest cover we now see in Australia has probably been in place for approximately 500 000 years (Archer et al. 1998, Martin 1987) (Figure 4.1). The arrival of Aborigines might have altered the vegetation in many parts of Australia to some degree (Dodson 1992); they might have partly reversed a general spread of forests in wetter areas that occurred as the climate warmed during the 10 000 years since the last glacial period (McIlroy 1990). It is likely that burning accentuated existing trends towards less rainforest and more fire-tolerant vegetation. But there can be no doubt that the arrival of Europeans was more significant than the activities of the Aborigines. Forests and woods were cleared throughout the country. The 43 million ha now remaining occupy about 5% of Australia's land area, but they represent only 62% of the resource standing in 1788 (Resource Assessment Commission 1992a).

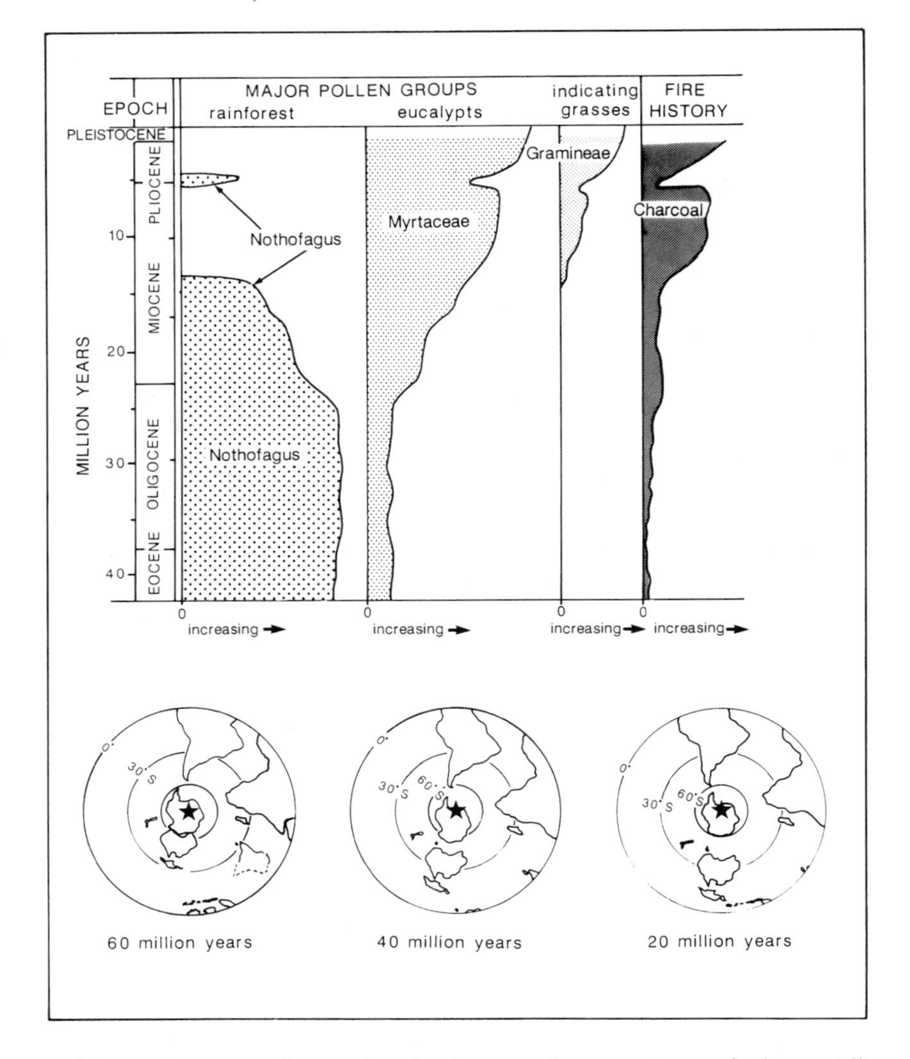

Figure 4.1
The decline of rainforest, and expansion of sclerophyll vegetation and grasslands with increased fire frequency, as Australia drifted northwards during the Tertiary. (Source: redrawn from Martin 1987.)

The main cause of extensive clearing was the necessity to feed a rapidly growing population and to develop an export market, especially in wool. A good forest was thought to indicate fertile soil, and was cleared for cultivation. But the early settlers were unimpressed by most of our trees as a source of timber. Trees were fire-damaged and hard to work, and timber warped when left to season. The soft red cedars in the coastal forests of New South Wales were godsends to the colonists. Nevertheless, the demands for fencing, housing, and fuel soon caused problems of supply. Within 20 years of settlement, regulations were enacted that forbade the cutting of trees along rivers in New South Wales, but these were rarely enforced and had little effect (Carron 1985). In South Australia in the 1850s agriculture spread north of Adelaide to supply food for the gold-rush population. Just beyond the perimeter of cultivated land was the Peachy Belt, a large area of peppermint gums. By 1880 this had almost disappeared, as its timber was plundered for fencing and firewood (Williams 1969).

Land-selection laws from the mid-1800s onwards required settlers to 'improve' their land by clearing. By the 1870s there was so much concern about indiscriminate clearing that timber reserves were set aside. In South Australia the *Forest and Trees Act* of 1873 gave financial incentives for landowners to plant trees. Two years later that State set up Australia's first forest service, which quickly began planting radiata pines to provide timber (Resource Assessment Commission 1992c). Although by the 1890s some two million ha were reserved for timber in Australia, there was little supporting policy, legislation or management. Drought, economic depression and political change at the end of the last century created a climate of such 'hostility to forestry [that] . . . in no other period was so much real damage done to forests', as one commentator wrote in 1921 (as quoted in Carron 1985, p. 7). Soldier settlements after the two world wars put further pressure on wooded lands, and clearing of land for agriculture continues today. In Victoria between 1972 and 1987 the rate of clearing on public land was almost balanced by the rate of reforestation. But on private lands the rate of reforestation was less than 20% of the rate of clearing.

Foresters thus saw themselves as the defenders of forests against the expansion of agriculture. In 1926 C. E. Lane-Poole, who became principal of the Australian Forestry School from 1927 until 1944, spoke vehemently on the matter:

> The Minister of Lands . . . is, as a rule, the type that is driven frantic by the sight of an un-subdivided region on the departmental maps. . . . the general public encourage the Minister . . . So he loudly joins in the cry of 'manless land for landless men'. To the Minister all land is potentially agricultural, all is too valuable to devote to forestry.
>
> As quoted in Powell 1988, p. 160

The years between the world wars were a significant period in the development of Australian forestry: it was a time of growing professionalism in the forestry services; coniferous plantations, and paper and pulp manufacture, were expanded; and increasing attention was given to the management of native forests. In New South Wales the 1935 *Forestry (Amendment) Act* set up well-secured forests for multiple uses: timber primarily, but also catchment protection, recreation, grazing, and wildlife conservation (Carron 1985). In these, as in forests elsewhere in Australia, the aim was to manage for sustainable yield. The concept of 'sustainable yield' came from European forestry and implied that forests would have approximately equal proportions of timber in all age classes that could supply a steady yield of timber. It was a concept that—like many other imports—did not fit the Australian environment well (Frawley 1987). Australian forests, when left unlogged, often have a high proportion of old trees, or are almost even-aged because of the effects of fire. Thus, many people feel that the aim of integrated and sustainable management was not achieved. Even if sustainable yield of production were achieved, sustainable use would require that ecological integrity also be achieved. Given that harvesting can affect wildlife, biodiversity, water quality and water balance in catchments, sustainable use is a difficult goal and it is easy in many cases to demonstrate that it has not been reached (Department of Environment Sport and Territories 1996). But to be fair, we should note that the theory of sustainable use in forestry anticipated the general adoption of such views by most environmentalists and the general public by some 30 years. It is not hard to see why L. T. Carron (1985, p. xiv) comments ruefully:

> The forestry profession [has] . . . an enormous record of achievement in spite of frequent political opposition and public apathy in the early days . . . [It] has been a major influence in ensuring that there is at least a forest estate in this country . . to haggle about.

Many factors contributed to the odium that fell on the forestry profession as the environmental movement gained strength. After the Second World War, demand for wood increased rapidly, as housing and industry expanded. Inventories suggested that supply could not meet future demand, and this was the basis for a vigorous program of softwood planting. In Tasmania pulpwood production served new paper mills, and demand from Japan in the 1960s encouraged rapid expansion. In New South Wales, near Eden, clearfelling of native forests to supply woodchips to Japan began in 1967 amid a storm of protest. Large areas (coupes up to 800 ha) were cleared along a major highway used by many people from Sydney, Melbourne and Canberra. The visual impact was severe, whereas most harvesting from native forests had previously been single-tree or small-group felling, which was unobtrusive (Carron 1985). Environmentalists vehemently challenged the operations on ecological and economic grounds. The new forest practices near Eden, and near Manjimup in Western Australia, were labelled a 'takeover of Australian forests for pines, woodchips and intensive forestry', and the 'Fight for the Forests' was on (Routley and Routley 1974).

Environmentalists also competed directly with the foresters for land, as pressure increased for more national parks to be created. Since the 1970s the names of once-unknown forest areas have become famous for the controversies that their management generated: Terania Creek, Daintree, Lemonthyme (Box 6). The controversies have been so widespread across the country, and so difficult to resolve, that the first inquiry commissioned under the *Resource Assessment Act 1989* looked into the forest and timber industry.

Box 6
Conflicts over forests in northern New South Wales

The forests in Australia are in the same regions as the land most sought after for agriculture: the land with good rainfall and a reasonable growing season (Figure 4.2). The moisture index shown in Figure 4.2 is the ratio of actual to potential evapotranspiration, and depends on both temperature and precipitation. At zero there is no moisture available for plant growth; at 1.0 there is a surplus of moisture for plant growth. Where the index is high in winter, land use options are greatest, particularly if the summer index is also high. Fruit and vegetables, tropical crops and sugar cane, dairying and intensive beef cattle raising are all potential competitors within the climatic zones suitable for most forestry. Historically, good forest growth was seen to indicate fertile soils, suitable for cultivation. At present not only agriculture but also nature conservation competes with forestry as the favoured land use. Thus, conflicts concerning the use of forested land are inevitable.

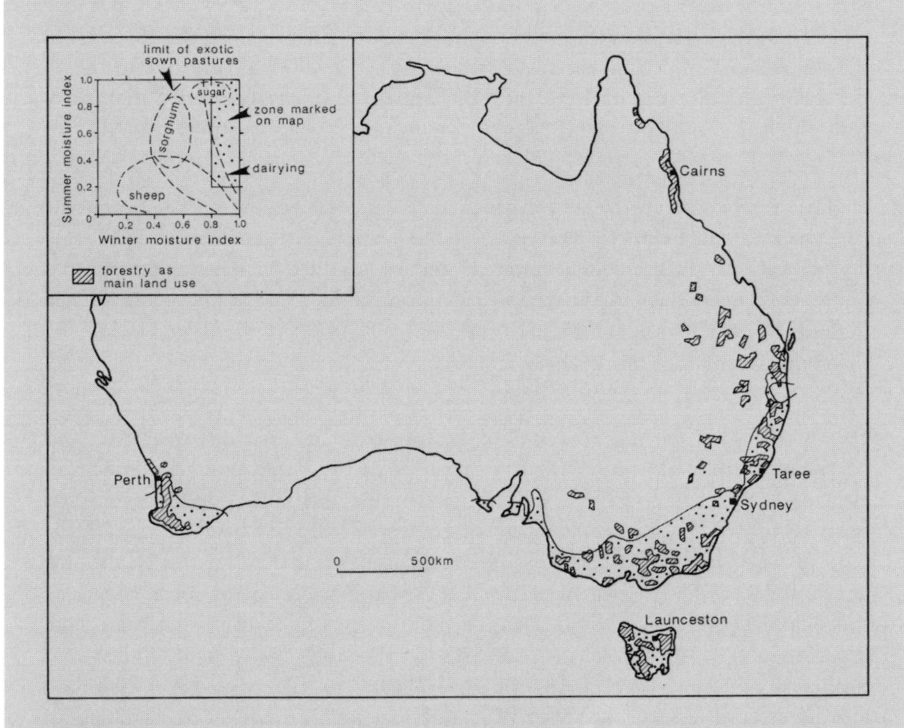

Figure 4.2
The distribution of forests in Australia, relative to climatic zones favourable to intensive agriculture. (Source: data from Division of National Mapping 1986.)

Politically, forestry has had little support except perhaps in the mid-twentieth century. In New South Wales the Forestry Department was described in parliamentary debate in 1892 as 'a child whom nobody owns', and when it found a home, that home was in the Department of Lands, whose primary responsibility was for land settlement. In the early twentieth century several factors encouraged the New South Wales Minister for Lands to open up forested Crown land for agriculture:

- Drought and economic depression in the 1890s had driven many small settlers off farms elsewhere, both from wheat farms in semi-arid areas, from sugar cane farms further north and from dairy farms on the New South Wales south coast.
- The newspapers strongly supported the opening up of land. In 1905, when the Minister for Lands visited Dorrigo in northern New South Wales and actively promoted its settlement, he ensured that the Sydney press accompanied him so that he would gain maximum political advantage from his decision.
- Changes to the dairying industry (development of refrigeration, invention of the cream separator, establishment of cooperative factories, pasteurisation, and widespread use of artificially sown grasses) combined to make dairying profitable on the previously remote north coast.

Even in the early days there were protests about the effect on the forests, but these focused as much on the waste of good timber felled during clearing as on the dwindling resource itself. A vocal but unheeded minority warned that it was folly to clear good forest land to provide second-class agricultural land.

> The destruction of the jungle which flanks the tablelands of Comboyne and Dorrigo should be strictly prohibited . . . The North Coast abounds in valuable timbers, belts of which should be reserved [and] . . . not tilled or cleared. Unfortunately in new country there is a tendency to skin it of all its wealth without any thought for the future. Flanks of tablelands and slopes of ranges are cleared and burnt with no thought of evil consequences.

So wrote H. I. Jensen, a government geologist, in his assessment of the soils of New South Wales in 1914. Photographs in his book show the effects of clearing and burning (Figure 4.3). Burnt stumps and fallen logs after 'a good burn' testify to the luxuriant nature of the forest that has been destroyed; the crop of waist-high grass surrounding an apparently satisfied farmer 12 months later indicates why the clearing seemed at first to be a profitable idea. However, this euphoria could not last, because the later crops of pasture could not match those that grew in the ash of the forest. Inevitably, many farms failed and timber production became the major industry in Dorrigo. This expanded rapidly after the Second World War, and regeneration of the trees failed to keep up with cutting. Introduction of chainsaws allowed quick felling, and both rainforest and blackbutt forest were overcut. Logging moved into less accessible mountain areas, as well as continuing in the forests on the coastal plain.

Figure 4.3
'A good burn' (A) and lush pasture growth one year later (B), Dorrigo district, New South Wales. (Source: Jensen 1914.)

Before the 1930s most production from the Border Range forests had been of hoop pine, but mechanisation led to utilisation of more species and greater areas. After the Second World War, quotas were set above sustainable capacity, which gave ammunition to the Forestry Commission's critics. The problem grew as sawmilling became economically more marginal. In the 1960s and 1970s people seeking an 'alternative' and 'back to nature' lifestyle moved to the north coast. They reacted strongly to forestry operations and also clashed with the existing population of timber workers, who saw themselves as practical conservationists. The protests against logging took on elements of a religious crusade.

In 1979 the blockade by protesters in the Big Scrub west of Lismore at Terania Creek—a small catchment of political, rather than natural, significance—began a long and bitter struggle that continues still. The conflict over forestry had shifted ground. Although there is still concern about clearing for agriculture, the main controversy is whether forests should be used for timber production, or reserved for conservation. Despite extensive policy changes, and the introduction of detailed environmental impact statements for logging operations, conservationists continue to reject the forestry operations as an acceptable use of the forest resource. Yet at least over the short term there is evidence of successful management. A blackbutt forest north of Taree was logged over a 30-year period, with non-merchantable trees being removed to encourage regeneration. Comparison of logged and unlogged plots showed that the volume of merchantable timber increased without any apparent effects on water quality or biodiversity. The Ecologically Sustainable Development Working Group comments that the risk of irreversible impacts on the environment in forests is much lower than for many other economic sectors.

Wiangaree, Roseberry and Mt Lindsay State Forests were declared in 1917 to protect the forest from clearing for agriculture. The Lamington National Park across the Queensland border was declared in 1915, and the Bunya Mountains was protected in 1908, despite the value of their timber, because the bunya pines were recognised as significant to Aboriginal culture, as scientifically interesting and as visually striking. Proposals to reserve the New South Wales side of the Border Range forests for conservation had been put forward in 1915 and were raised again in the late-1940s. But it was not until 1983 that large tracts of State forest were given national park status. In 1986 their protection was reinforced when 15 national parks and nature reserves became Australia's sixth World Heritage property.

Data from Carron 1985, Ecologically Sustainable Development Working Groups 1991, Frawley 1987 and 1988, Jensen 1914, Watson 1990

What are our forest resources?

About 156 million ha of land in Australia are covered by native woodland or forest (that is, have a tree cover with existing or potential canopy closure of 20% or more). Of this total area, 42 million ha are commercial forest, and it is this lower figure which is the basis of many forest resource assessments, for example those by the Resource Assessment Commission. Plantation forests (86% exotic pines and 14% native, mainly eucalypts) occupy only another one million ha, but there was an intergovernmental agreement in 1997 to expand this to three million ha by 2020 (Agriculture, Fisheries and Forestry—Australia 1999).

As with any resource, an assessment of forest resources must begin with caveats about the limitations of data. The comprehensive national survey of commercial forests, produced by the Resource Assessment Commission (1992d), comments that 859 different types of forests and woodlands were used by forest services in Australia when the services replied to a questionnaire sent out by the Commission. It is hard to see how we can compare data collected when the classification systems used by different forest services are so diverse. Furthermore, there are the difficulties of estimating loss of vegetation types, even on the most basic classification of 'forest' and 'woodland', as discussed in Chapter 2. The Resource Assessment Commission aggregated the 859 types into 15 groups for the Forest and Timber Inquiry, as a basis for further analysis (Table 4.1). The tenure under which these groups were held was listed in four categories:

- state forest (multiple-use public forest reserved primarily for timber production)
- crown land (other public forest in which timber harvesting may occur)
- conservation reserves (public forest reserved primarily for conservation and/or recreation)
- private forest (all privately owned forest).

Australia-wide, there are approximately equal areas (of 13–17 million ha) held in state forest, crown land and conservation reserve. Apart from the conservation reserves, about 60% of each tenure is available for logging. The remainder is inaccessible, uneconomic to develop at present, or reserved for conservation purposes (Resource Assessment Commission 1992b). There are wood production regions in the eastern States and Tasmania, but large parts of the central and western States have no economically significant woodland or forest. Leasehold land, usually pastoral leases which are wooded and which require governmental permission to clear, occupy a further 66 million ha (ABS 1999).

Less than 30% of our forests are privately owned. This contrasts with other developed countries, such as Germany, Japan and the USA, where the corresponding figure is about 70%. Clearly this fact accentuates the importance of government actions and policies in controversies over the use of forests. Approximately 1% of the Australian workforce is employed directly in forestry, milling and wood processing (Ecologically Sustainable Development Working Groups 1991b); the value added by these industries was about 1% of Gross Domestic Product in 1988–89. However, imports of wood products, particularly pulp and paper, still far exceed exports (Resource Assessment Commission 1992a). The forest industry is an important industry, but one with economic as well as environmental questions hanging over it.

From Table 4.1 it is clear that Australian forests and woodlands, while dominated by eucalypts, are nevertheless very diverse. There is a clear productivity gradient that is climatically controlled. Wet sclerophyll forests (groups 7 and 8) in south-eastern Australia support over 300 tonnes/ha biomass; dry sclerophyll forests (group 10) support less than 200 tonnes/ha; poplar box woodland (group 15) supports about 150 tonnes/ha; and in areas of mallee woodland, the figure falls below 50 tonnes/ha (Resource Assessment Commission 1992c). In forestry, productivity is measured by two indices: merchantable wood volume (MWV, m^3/ha) estimates the standing resource, and mean annual increment (MAI, m^3/ha/year) estimates the growth rate. Variations in these indices for forests and plantations are shown in Table 4.2.

Table 4.1
Native forests and woodlands in Australia

	Forest group	*Locations*	*Tenure*	*Total area (ha)*
1	tropical rainforest	Qld, NSW, NT	almost entirely held in conservation reserves	1 146 000
2	subtropical rainforest	coastal NSW, Qld	in Qld, mainly held in State forest; in NSW, mainly in conservation reserves	396 000
3	temperate rainforest	NSW, Vic., Tas.	mostly in State forests	1 002 000
4	mangrove and swamp forest	NT, Qld, NSW, Vic.	all held in conservation reserves, except in NT	913 000
5	SW wet eucalypt forest	WA	in State forests and conservation reserves	184 000
6	SW dry eucalypt forest	WA	over 50% in State forests	2 220 000
7	SE wet eucalypt forest	NSW, Vic., Tas.	in Vic., mainly in conservation reserves; elsewhere mainly in State or private forest	4 064 000
8	SE ash forest	NSW, Vic., Tas.	in NSW, mainly in conservation reserves; in Tas. and Vic., in State forests and conservation reserves	1 482 000
9	SE dry forest and woodland	SA, Vic., NSW	in SA and Vic., mainly State forest; in NSW, private land and Crown land	6 429 000
10	SE coastal eucalypt forest	Vic., NSW	in Vic., mainly State forest; in NSW, mixed tenure	3 775 000
11	central coastal eucalypt forest	NSW, Qld	State forest and conservation reserves	4 241 000
12	NE coastal eucalypt forest	Qld, northern NSW	in NSW, conservation reserves; in Qld, mixed tenure	2 738 000
13	river red gum forests	Qld, NSW, Vic.	in Qld and Vic., mainly conservation reserves; in NSW, mixed tenure	456 000
14	native pine forest and woodland	NT, Qld, NSW, Vic.	near coast, mainly conservation reserves; inland, State forest and Crown land	4 018 000
15	northern dry forest and woodland	NT, Qld, NSW	in NT, mixed tenure; in Qld conservation reserve and Crown land; in NSW, conservation reserves	10 119 000
	Total area of forested and wooded lands			43 185 000

Data from Resource Assessment Commission 1992d

Table 4.2
Productivity of Australian State forests and plantations, by type*

Group	*NSW*	*Vic.*	*Qld*	*SA*	*WA*	*Tas.*
NATIVE FORESTS						
tropical rainforest			4–40 (0.2–0.3)			
subtropical rainforest			6–30 (0.3)			
temperate rainforest						77 (2.0)
SW wet eucalypt forest					150–200 (8)	
SW dry eucalypt forest					34–135 (0.8–1.6)	
SE wet eucalypt forest	20–75 (0.5–3)	30–235 (1.4–1.9)				145 (2.5)
SE ash forest 249–325	100–130 (2.5–4)	70–360 (1.9–4)				(5.1)
SE dry forest and woodland	10–45 (0.3–1)	10–20				
SE coastal eucalypt forest	25–120 (0.5–3)	9–195 (0.6–1.5)				
NE coastal eucalypt forest			1–30 (0.1–0.2)			
river red gum forest	35 (0.5)	2				
native pine forest and woodland	10 (0.4)		5–15 (0.1–0.2)			
northern dry forest and woodland			0–15 (0.2)			
SOFTWOOD PLANTATIONS						
Pinus elliotti			150–200 (9–12.5)			
Pinus caribaea			200 (11.5–13.5)			
Pinus radiata	150–255 (15–16.5)	160–185 (18)		(15–20)		<280 (15.5)

Data from Resource Assessment Commission 1992d
**Measured as MWV (merchantable wood volume in cubic metres per hectare).*
Figures in parentheses represent MAI (mean annual increment in cubic metres per hectare per year).
Single figures denote mean (average) values rather than ranges.

The productivity of the ash forests of the south-east increases as we move southwards from New South Wales through Victoria to Tasmania, illustrating the effect of increasing wetness. The controversial rainforests of Queensland have only moderate productivity, and this fact led many conservationists to argue that they were economically marginal and should be used for conservation, not timber production. Note also the high productivity of softwood plantations in comparison to most native forest. Among conservationists, but also among some farmers and within the general community, there is long-standing prejudice against pine plantations, but they are certainly valuable sources of timber. For example, the dry forests and woodlands have mean annual increments of less than 1 m^3/ha, an order of magnitude below the productivity of the pine plantations that may replace them. Productivity is also increased by more frequent logging of the plantations. Logging rotations for sawlogs in native forests are long (less than 80 years in New South Wales, Victoria and Tasmania, and about 40 years in Queensland's dry forests and woodlands), and are considerably shorter in plantations (Resource Assessment Commission 1992d).

Timber, including sawlogs and pulpwood, is not the only resource derived from forests. Fuelwood production, honey production, grazing, seed production, recreation and tourism, flora and fauna conservation, and catchment protection are also forest industries. But economically these are comparatively minor industries and, apart from grazing, are not perceived to have significant environmental impacts. It has been the balances between sawlog and pulpwood production, and between wood production and conservation, that have been the focuses of most controversies.

Conservation and old-growth forests

Debates over forest management tend to be highly emotive. Only controversies over large dams are as likely to generate television footage of protesters blocking the roads in front of trucks and bulldozers, and being dragged from the scene. The language used is often forceful. J. Formby (1991, p. 6) criticised the Resource Assessment Commission for not explicitly questioning the forest agencies and their 'timber-production ethos, arrogance, lack of public accountability, inadequate scientific research and other failings'. Perhaps more than any other issue, the debate over Australian forests has become polarised, and characterised by mistrust and aggression between the conflicting groups. The Resource Assessment Commission inquiry yielded much information about forests in Australia, but brought about little resolution of these fundamental and politically significant differences. The National Forest Policy Statement in 1995 was an attempt to resolve the controversies, by establishing a procedure for regionally based, well-researched, widely consultative agreements that met conservation goals but ensured resource security and jobs. The States are involved because of their constitutional responsibility for resources. The Commonwealth is involved because of its obligations in World Heritage areas, under heritage legislation and via other international agreements such as the Convention on Biological Diversity and the Global Statement of Principles on

Forests (Dargavel 1998). Several Regional Forest Agreements have been made but the process has encountered delays and criticism. Kirkpatrick (1998) argues that the Tasmanian agreement does not protect reserves adequately from development; Horwitz and Calver (1998) question the lack of peer review of scientific assessments in Western Australia; Hutton and Connors (1999, p. 257) comment that 'the one state where the process seemed to work well was Queensland'. Dargavel (1998, p. 30) concludes that they cannot be evaluated yet and will probably show 'both new strengths and old flaws'.

Despite the well-documented clearing for agriculture, it is possible to argue that Australia's forested areas are now well protected in reserves. Only 5% of Australia's land is still forested, but forests and woodlands cover 25% of Australia's national parks and conservation reserves. Of our total wooded area, 11.3% is in reserve, a figure exceeding the IUCN/WWF global criterion of 10% of current forest cover (Agriculture, Fisheries and Forestry—Australia 1999). Of the land-based World Heritage areas, only Uluru and Willandra Lakes are largely treeless. The Lord Howe Island Group, East Coast Rainforest Parks, Queensland Wet Tropics and Kakadu are more than 75% forested, and the Western Tasmanian Wilderness is about 50% forest, with the rest being largely moorland. In areas listed as World Heritage, and also in areas being considered for listing, the Commonwealth government has the responsibility to intervene to forbid any activities that may detrimentally affect the area's heritage values. Additionally, other forested areas are listed on the Register of the National Estate. Contrary to many people's belief, this is not a direct protection, but it does give credibility to calls for conservation and sensitive management (Resource Assessment Commission 1992b). The Ecologically Sustainable Development Working Group (1991, p. 29) comments that 'by any conceivable yardstick, forest use and management compares favourably with the environmental impacts of the agricultural sector'. Perhaps the forests are now relatively well conserved, simply because they occupy land that was less favoured for agriculture, or because they were seen as compatible with grazing, and therefore they were available when the political climate became right. But this is not the whole story; the demand for conservation in Australian forests echoes a worldwide trend that values forests, particularly rainforests, above other vegetation communities. The vegetation communities that occupy most of Australia simply do not have the same allure.

Conversely, we could argue that our forests are a dwindling resource and that their scarcity is a powerful reason to preserve them. Our few remnants of rainforest are—in the words of the World Heritage Convention—'outstanding examples representing significant on-going ecological and biological processes in the evolution and development of . . . ecosystems and communities' and 'significant natural habitats for in-situ conservation of biodiversity'. Australia's rainforests have been cleared extensively, and now their protection is well justified. For example, K. Mills (1988) argues that the common belief that the Illawarra region of New South Wales was covered by rainforest at the time of settlement is incorrect. Natural grasslands, and sclerophyll forest and woodlands, were the main vegetation types. Nevertheless, after a careful reconstruction based on historical records and ecological interpretation, he concludes that, of an original 23 000 ha, only 5800 ha remain. This represents only about 25% of the pre-European stands. However, in north Queensland approximately 80% of the original stands remain (Frawley 1987), and this reinforces the value of the areas now protected by the World Heritage legislation.

Also, continent-wide statistics hide the diversity of the forests (see Table 4.1) and give no indication of the significance of habitats within them. The 'precautionary principle' of ecologically sustainable development also applies (Ecologically Sustainable Development Working Groups 1991). The Intergovernmental Agreement on the Environment (IGAE 1992) defines this principle as the responsibility to take measures to prevent degradation, and to avoid serious or irreversible damage to the environment, even if there is not complete scientific evidence to show that damage will certainly occur. However, the present system of conservation reserves is not ideal and does not achieve the aims of the precautionary principle. A study in Tasmania looked at reserves using two methods: vegetation mapping and environmental domain analysis. Both showed that the existing reserve system was not representative of the full range of ecosystems. No extinctions of native flora or fauna due to timber harvesting in Australia are known, but our knowledge about forests is too limited to be sure that none have occurred (Resource Assessment Commission 1992a). In response to these concerns, one of the goals of the National Forest Policy is a 'comprehensive, adequate and representative' reserve system. This aims to have all forest types well represented in reserves, a goal still far from achievement (Department of Environment Sport and Territories 1996). Specifically the goals under Regional Forest Agreements are 15% of all forest ecosystems that existed in 1788, plus 60% of existing old-growth forest and 90% of high quality wilderness, in reserves.

Calls for higher levels of conservation often focus on old-growth forests. Old-growth forests have been variously defined, but the term includes forests that:

- have been disturbed relatively infrequently or to a minor degree since settlement (that is, they are mainly unlogged)
- are ecologically mature or, in forestry terms, 'over-mature', and probably have high structural diversity.

Old-growth forests are the most valuable forests for wildlife and flora, and often the most visually attractive. They also seem to arouse the strongest sense of spiritual association with nature. Both the overstorey trees and the understorey can be hundreds of years old. In the mountain ash forest north-east of Melbourne, specimens of tree ferns (*Cyathea* and *Dicksonia*) and understorey trees (*Olearia* and *Persoonia*) yielded radiocarbon ages of 98–370 years, and the *Eucalyptus regnans* of the overstorey may live 350–450 years (Mueck et al. 1996).

But it is not only in old-growth forests that concerns about species extinction have been raised: it is necessary to conserve wildlife not only in reserves, but also in forest that is logged. Clearfelling, rather than selective logging, has become far more widespread over the last 20 years. Animals that forage over large areas, rely on hollows in old trees for shelter, or are 'central place foragers' suffer under this practice. 'Central place foragers' are the larger and more social animals and birds. Because they search for food in all directions from their nests or homes, they are disadvantaged in the narrow linear creek reserves that provide most of the refuges in clearfelled areas (Recher et al. 1987).

Wildlife conservation in forests

The Resource Assessment Commission (1992b) considered two questions relevant to wildlife conservation in forests:

- Does logging increase the risk of species extinction and hence decrease forest biodiversity?
- Are current practices designed to reduce impacts of wood-production activity adequate?

It gave cautious answers, stressing the lack of available information, but found no immediate threat or evidence of major ecological damage on a regional scale. Victorian studies suggested that wildlife corridors in logged areas could be adequate to protect gliders and possums. The Commission examined claims that foresters had breached management codes designed to protect ecological values, but found that breaches were uncommon. Again, these conclusions may be valid on a continent-wide and general level, but they certainly do not convince all environmentally aware people that forestry has no adverse effects. Forests are still fought over, and the debate has become so polarised that wood production and conservation are seen as incompatible (Box 7). This is unfortunate, because examples of successful integrated management exist, such as the state forests at Kioloa (New South Wales) and Perup (WA) (Davey and Norton 1990).

The South-east Forest Region of New South Wales is the area that provides woodchips for export from Eden (Plate G). The forests were not considered to be of prime quality and were severely affected by wildfires, especially in 1939 and 1952. There were dense and even-aged stands in many places, as well as areas where regeneration after the fires had been very poor; uncontrolled logging had further degraded the resource. In forestry terms, woodchip production represented a sensible use of the resource (Carron 1985). Environmentalists saw the matter differently. They attacked the economics of the project and also its environmental consequences:

> despite the fact that drastic modifications will be made to an enormous area, . . . no biological survey has been undertaken or is contemplated. . . . the refuges [for wildlife], though not worthless, will be of very limited value . . . [and] strips [of vegetation] retained in their natural condition, allegedly for the purpose of wildlife conservation . . . will be burnt very frequently, thus greatly diminishing their value.
>
> Routley and Routley 1974, p. 103

The Forestry Commission responded to these criticisms, reducing the size of coupes dramatically, and increasing the refuge areas. Considerable research into the distribution and habitat requirements of native fauna has been carried out (Pyke and O'Connor 1991). Yet there remains much disagreement and controversy, with conflicting calls for logging to protect the woodchip industry, and for conservation to protect wildlife and vegetation. D. Mercer (1995, p. 79) is critical:

> So large is the commitment to the massive annual timber cut in the Harris–Daishowa concession that essential wildlife corridors along streams have been considered an unjustifiable 'luxury' and so have been progressively alienated since 1977 when earlier logging prescriptions were set down.

Recher et al. (1987, p. 191), discussing guidelines based on the Australian Museum's wildlife research, saw the matter differently, in the late 1980s:

> In addition to retaining mature forest along creeks and gullies and providing movement corridors between catchments and large reserves (e.g. national parks, flora reserves), as is currently practised at Eden . . . large patches of mature forest need to be retained within the logging area.

The forest industry believes it has a strong commitment to faunal conservation. In the south-east forests, the total area of State forest is 245 000 ha, and there are 155 000 ha of national parks/nature reserves. *The Environmental Impact Statement* (State Forests of New South Wales 1994) focuses strongly on fauna protection. Logging is excluded from 29 000 ha in the State forests. To maintain production of pulpwood from the forests in the Eden area for a period of three years, a 'Least Sensitive Area' of 60 000 ha is identified in which normal logging operations would occur. Within this area, filter strips along creeks, and wildlife corridors, are maintained; clearfelling is not practised, and coupes are logged alternately to produce a mosaic of logged and unlogged forest. Rotation will be short, with coupes being logged on a 20–40 year cycle. However, areas associated with animals of high conservation significance—such as the koala, large forest owls, the long-footed potoroo and the southern brown bandicoot—are excluded from the short-term logging program; a moratorium to allow further investigation of habitat significance is placed on these areas, which total 89 000 ha.

Conservation practice in forests aims to maintain viable populations of all species, and to give special protection to rare and threatened species. It is based on the linking of refuge areas, along drainage lines and other wildlife corridors; the logged areas have limited conservation value. However, it is not easy to zone or manage forests to include the full range of habitats and to prevent degradation by weed and feral animal invasion as the forest becomes fragmented. Furthermore, habitats, like the animals, are not static. The habitats of nine arboreal marsupials in the Kioloa area were identified using a wide range of environmental variables. Two factors described them best: one related to soil and leaf phosphorus levels, and the other to moisture and/or leaf nitrogen levels (Figure 4.4A). As these factors could be related to forest types, the expected populations for each animal could be estimated. Also, habitats expanded during wet years and contracted during drought (Figure 4.4B). As both drought and fire accentuate the loss of populations due to logging, managing forests for wildlife conservation has to take account of climatic variation and fire regimes (Davey and Norton 1990). G. H. Pyke and P. J. O'Connor

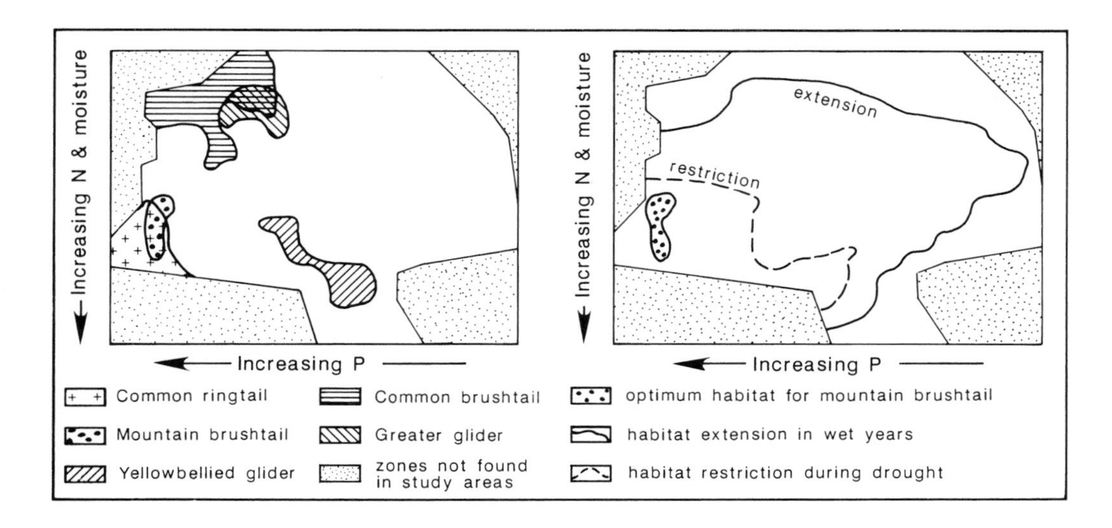

Figure 4.4
Diagrams showing:
A the differing sites—defined by phosphorus levels, and nitrogen and moisture levels—that are preferred by arboreal marsupials in the south-eastern forests of New South Wales
B the expansion and contraction of the habitat of one marsupial, the mountain brushtail, in response to wet years and drought.
(Source: redrawn from Davey and Norton 1990.)

(1991) challenge the 4–7-year cycle, autumn and winter burn, low intensity fire regime used in the south-east forests. Some plants take longer than seven years to reach flowering maturity; others need a hot fire to germinate. Some animals probably evolved under a summer fire regime and may be ill-suited to a changed fire season; others thrive in the conditions that occur five or more years after a fire. They recommend a change to a 10–25-year cycle, with occasional hot summer burns, and a diverse fire regime throughout the area.

There is particular concern for predators near the top of the food chain, such as the spotted-tail quoll and the powerful owl, that need a large home range. The owls feed on animals whose abundance is likely to drop as a result of logging. Because they live for several decades, any effects of past forestry practices may remain undetected for some time. Also, while many animals may return within a 30–40-year period after logging, those that need tree hollows may be affected permanently. The logging rotation for sawlogs is 80–120 years, yet tree hollows seem to develop only in older eucalypts (more than 200 years old). This means that retaining habitat trees in logged areas is vital to wildlife conservation (Pyke and O'Connor 1991). In the montane ash forests of Victoria, where a 50–120-year cutting cycle is used, D. Lindenmayer (1994) asserts that clearfelling is a serious threat to arboreal marsupials such as the rare Leadbeater's possum. He advocates a move from clearfelling to retaining more vegetation on logged sites, the provision of wildlife corridors in logged areas, and a larger area of national parks.

Other critics challenge the basis of reservation strategies. The linking of physical parameters, such as soil type and local climate, to vegetation (as in Davey and Norton's analysis) is called the 'environmental envelope' strategy. D. McIlroy (1990) criticises this method for identifying areas worthy of reservation. For example, he notes that in the Wet Tropics World Heritage Area of Queensland the boundaries between even the sharply differentiated vegetation types of savanna and rainforest are not explained by physical parameters, because of variations in fire histories and in the pattern of historical land-use changes.

Box 7
Woodchips in Tasmania

Tasmania is Australia's most forested State, and has had the most publicised conflicts between development and conservation. It contains the last areas of wilderness with cool temperate rainforest in Australia, and after much controversy, these are now protected by World Heritage status. However, most of Tasmania's forests are not rainforest. About 85% are eucalypt forests, and the use of these is an important environmental consideration in Tasmania. Plantations are also significant (Plate H).

In 1991 about one-quarter of Tasmania's area was under the control of its Forestry Commission. Tasmania led the other States in production of pulpwood and paper from its native forests, and the first mill began operations at Burnie in 1937. Other mills followed on the Derwent River at Boyer in 1941, and near Hobart at Geeveston and near Devonport at Wesley Vale in 1961. In the late 1950s companies were given rights to take pulpwood from forests in south-west Tasmania also. The industry grew slowly until the 1960s, when Japan became an export market, and export began from Longreach near Wesley Vale and from Triabunna near Hobart. In 1988 construction of a fourth plant at Whale Point near Hobart was approved.

In the five years after 1969–70 financial year, sawlog production from Crown lands remained steady, but pulpwood production increased and pulpwood export began. Sawlog production from private lands declined, but pulpwood production increased. The Forestry Commission saw export of woodchips as a welcome opportunity; opponents saw it as the cause of forest devastation. By the 1980s the situation was complicated further as sawloggers sought equal assurance of supply as the pulpwood and paper industries enjoyed under the concession system. There were deep divisions in the community, mirroring those caused by the controversies over the Gordon-below-Franklin issue. The Commonwealth government courted the environmental vote and viewed woodchipping unfavourably, but the Tasmanian government and trade union council supported woodchipping.

Management dilemmas also arose. The highland forests are a complex mosaic of temperate rainforest (dominated by *Nothofagus* and *Phyllocladus*), multiaged eucalypt forest, and open *Poa* grassland. The forests dominated by *Eucalyptus delegatensis* had an open grassy floor, probably because of Aboriginal burning, and in the absence of fire, there would be a succession from grassland to forest to rainforest over a period of 100–150 years. Fire has the effect of reversing this process, so that even before European settlement, old-growth rainforest was confined to protected, valley sites. Since settlement all three types have changed in some places, with rainforest or other shrub storey replacing grass under eucalypts, and with rainforest being burnt out. Some logging of the eucalypt forest was carried out up until the 1960s, and in the 1970s areas were clearfelled for pulpwood. Since then a rainforest understorey has developed, and regeneration of eucalypts has not been satisfactory everywhere. This presents a management problem for foresters, as the rainforest understorey is associated with dieback of the eucalypts, perhaps because it encourages microbial growth in the soil. Grazing and burning have both been discouraged, yet these would probably reduce the rainforest understorey. The central problem in making management decisions is the difference in the objectives of the Forestry Commission, the pulp and paper companies, and private owners. Elsewhere, however, the aims of these groups are often in accord, with all parties regarding the development of woodchip operations as beneficial.

By 1988, three companies—Australian Pulp and Paper Mills (APPM), Tasmanian Pulp and Forest Holdings (TPFH), and Forest Resources—had licences to export 2 825 000 tonnes of woodchips from Tasmania. An Environmental Impact Statement was required under the Commonwealth's *1974 Environmental Protection (Impact of Proposals) Act* when extensions of the licences beyond 1988 were sought. But disagreement between the State and Commonwealth governments about areas listed as National Estate led in 1987 to the setting up of the Inquiry into the Lemonthyme and Southern Forests to decide if these included areas qualifying for World Heritage listing. This Inquiry—known as the Helsham Inquiry, after its chairperson—was the eleventh to be held in Tasmania on forest issues within a 12-year period, and this fact illustrates the significance of forest controversies both in Tasmania and nationally.

The Inquiry found that areas qualifying for World Heritage listing did exist, and that reserving them would have only a minor effect on the forest industry. At the same time, controversy was raging over APPM's proposal for a world scale kraft pulp mill at Wesley Vale. While the Commonwealth Minister for Industry hailed this as the biggest private investment in manufacturing in Australia, conservationists opposed it on the grounds that the chlorine bleaching process would lead to discharge of water polluted by dioxins into Bass Strait. They were supported by the CSIRO, by farmers from the surrounding rich agricultural area, who were concerned also about air pollution, and by the State's Department of Sea Fisheries. In mid-1989, after the Commonwealth set stringent environmental controls, the major foreign partner in the mill venture withdrew, claiming that the proposed standards were not attainable, and the proposal lapsed.

The findings of the Helsham Inquiry subsequently came under fire, with scientific witnesses challenging them. The conservation movement mounted a legal challenge, while the Tasmanian Government sought compensation from the Commonwealth for loss of resources, and the timber industry protested that two of the best sawlog areas had been alienated to World Heritage. In September 1989, largely because of the electoral success of the Greens in Tasmania, the South West Wilderness area under World Heritage protection was expanded by more than twice the area recommended by the Helsham Inquiry.

This was by no means the end of the matter. In January 1995 a licence to export woodchips was set aside after conservationists mounted a successful legal challenge on the grounds that the federal Minister for Resources had not considered the environmental issues adequately. The ruling placed other export licences in jeopardy and renewed calls to ban woodchip production in all native forests in Australia.

Data from Ellis and Lockett 1991, Ellis and Thomas 1988, Formby 1987, Mercer 1995, Van Saane and Gordon 1991

A question of language

Is clearfelling really such a drastic and unprecedented change to forest ecology? Are the conservationists exaggerating the problem? The pulpwood industry has argued that they are. Eucalypt forests have evolved under a regime of severe wildfires, which can almost completely remove all the vegetation and leave the hillslopes bare and exposed to erosion. Supporters of woodchip operations draw an analogy between the effects of wildfires and the effects of clearfelling, pointing out that both the natural events and the forest operations lead to regrowth forest. Forestry, they say, is simply replicating a natural process (Figure 4.5).

Perhaps the most interesting aspect of the analogy represented in Figure 4.5 is its use of language. Both causes—'fire, a natural occurrence' and 'planned forest logging'—lead to the same end—'regrowth forest'. The natural pathway is 'uncontrolled', 'baring the soil', causing 'destruction' and the 'death of trees under intense heat'. All these words convey images of devastation by natural

Figure 4.5 Flow chart suggesting that the ecological effects of clearfelling parallel those of wildfires. (Source: New South Wales Pulp and Paper Industry 1989.)

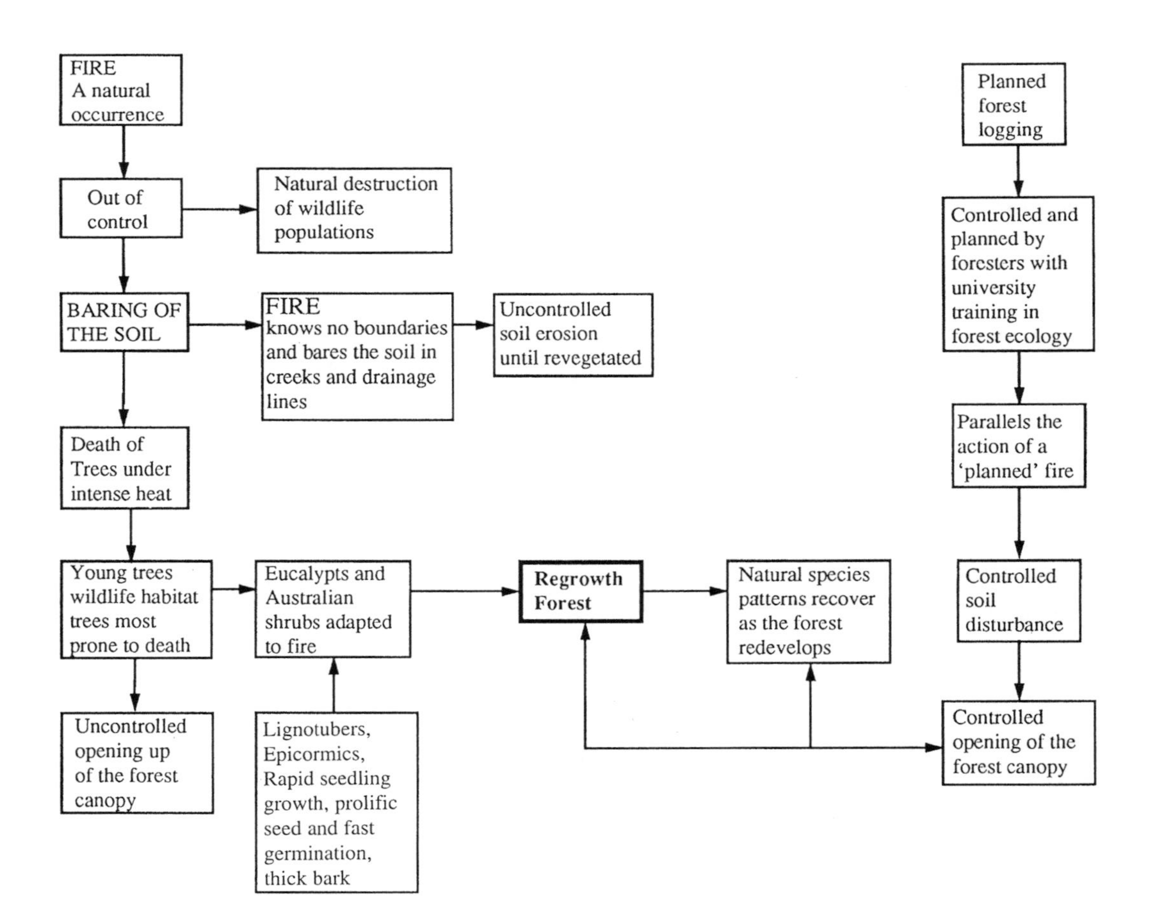

forces beyond our control; they arouse fear and imply that the effects may last for a very long time. The other pathway—described as 'logging' or 'harvesting'—is 'planned' and 'controlled' by experts, so that 'natural species patterns recover'. The images conveyed are of sensible and acceptable use, and imply that recovery will be rapid and erosion will not be severe. This impression is reinforced by the layout of the flow chart. Wildfires have side-effects: 'destruction of wildlife' and the spread of fire into creeks. No such side-effects are shown on the side of the flow chart representing forestry operations; on that side, the consequences of 'logging' are shown to flow simply and smoothly, reinforcing the impression of control.

This diagram is a clear example of the way in which language and layout can be used to create an impression favouring one side of an environmental dispute. In most disputes, both sides use analogies and language that reinforce their point of view. In weighing up an environmental argument it is important to look at the use of language, paying particular attention to analogy and metaphor, and to terms that are implicitly derogatory or flattering. Strong language may obscure a very weak argument. In this case, for example, the analogy is not valid. Clearfelling occurs over long periods, rather than the few days a wildfire takes to pass. Felling occurs regularly and it affects the whole forest, whereas fire damage is patchy. After a wildfire many trees resprout by epicormic growth, but in clearfelling whole trees are removed. Wildfire does not involve the movement of machinery and the compaction of soil that this causes. Forestry may cause more soil disturbance and remove more nutrients than wildfires.

However, in rejecting the analogy of clearfelling and wildfire, it is important to remember that inaccurate and emotive language is also used by opponents of forestry. A recent textbook on the Australian environment (Aplin 1998, p. 290) describes clearfelling as 'the equivalent of an army's scorched earth policy, cutting all timber . . . [so] that nothing else escapes devastation'. Selective harvesting which needs 'a more careful approach to harvesting timber' is the only other method mentioned. This stark contrast would have been well justified in Eden in the 1970s but takes no account of changes made by (or forced upon) forestry operations since then. We also need to discuss the question of wildfires further. Wildfires are as great a threat to forest production as they can be to wildlife and to property. Controlled burning is a deliberate strategy that is central to forest management and aimed at trying to reduce the hazard of wildfires. Yet what foresters see as a necessary control strategy, some conservationists see as wilful destruction of natural values.

A burning question

Australia has been called the 'fire continent' (Pyne 1991). European settlers used fire widely, but the power and ferocity of wildfires in Australia took them by surprise. In Henry Lawson's short stories, fires are represented as menacing and evil, emanating from a cruel and inexplicable landscape and from the hands of brutish and irresponsible people. On Black Thursday—6 February 1851—bushfires engulfed most of Victoria, and the terror of the scene is vividly portrayed in William Strutt's painting based on his own experience and accounts of survivors. It shows a chaotic flight of people and animals, all wide-eyed and terrified, rushing over the carcasses of birds, kangaroos and stock (Gleeson 1976). The effects of fires on forests were obvious. Baron von Mueller, the well-known and influential pioneering botanist, commented in 1882 that the best of the Western Australian forests suffered wholesale destruction of their best timber because of fire. Forest reserves were frequently burnt out by fires that spread into them from fires lit by settlers to clear land or encourage grass growth. In the huge conflagrations of Black Sunday—14 February 1926—in Victoria and the massive Black Friday wildfires of 13 January 1939, most of the initial fires were lit intentionally, but casually and carelessly, by farmers, hunters, loggers, foresters, campers and arsonists, in conditions under which control was impossible. Drought, hot dry northerly winds and following southerlies fanned fires on Black Friday that engulfed South Australia, Victoria, Tasmania and the ACT: 'For mile upon mile, the former forest monarchs were laid in confusion, burnt, torn from the earth, and piled one upon another as matches strewn by a giant hand' (Royal Commissioner Stretton, as quoted by Pyne 1991, p. 311).

Foresters wanted to control fire but differed about how to do it. Some, led by C. E. Lane-Poole, wanted to exclude fire and allow natural ecological succession to shift the forest to a less flammable state, resembling European forests. Others, particularly in Western Australia, favoured cleaning up the forest floor and setting frequent fires to reduce fuel (Figure 4.6). At first, fuel reduction was done within sawlog compartments that were also protected by firebreaks, but by the 1950s fuel-reduction burning throughout the jarrah forest was adopted (Bartle et al. 1981). A scientific basis for fuel-reduction burning began to develop with Foley's study of the meteorology that promoted fires, and R. H. Luke's outline of fire control principles. Even the outbreak of severe fires in the jarrah forests in 1960–61 did not shake the support for hazard-reduction burning. Foresters argued that the fires were exceptional, that no lives had been lost and that devastation would have been worse if fuel loads had not been kept relatively low by previous control burning. In 1962 a forest researcher, A. G. McArthur, published his account of the technique of control burning in eucalypt forests (Pyne 1991). Since then the semicircular Fire Danger signs, adjusted according to the meter devised by McArthur, have become a familiar sight on roads near forests and forested national parks. The meter is based on a nomogram that allows fire danger to be read from 'low' to 'extreme' on a 100-point scale, where 100 equals the conditions of the 1939 fires. It takes into account the number of days since rain, the rainfall to 9 a.m., relative humidity, air temperature and wind speed. If fuel loads are known, then the rate of spread of a fire, flame height and average spotting distance can be estimated. More recent developments in fire prediction involve complex computer models of fire behaviour.

Ironically, as confidence increased in the usefulness of controlled burning to reduce the impact of wildfires and in managers' ability to predict its behaviour, so too did opposition to its use. In the 1970s many forested areas were transferred from State forests to national parks. The Australian Conservation Foundation challenged the practice of controlled burning, emphasising the adverse effects of erosion, nutrient loss and disturbance of flora and fauna.

Figure 4.6
A 'clean' forest floor after a fuel-reduction fire.

Forest managers continued to support the practice. In the forests of south-west Tasmania, after devastating fires further east in 1967, preventive burning of button-grass moorlands was added to the existing practice of using hot slash fires to stimulate regeneration on logging sites. Thus, controversy over fire use took its place alongside controversy over dam construction and woodchipping in the south-west (Pyne 1991).

The current situation is perhaps best described as a stalemate. The Resource Assessment Commission (1992b) accepted that fuel-reduction burning is effective in reducing the hazard of wildfire and should continue to be used. Yet assessing its ecological impacts is very difficult. Two types of fires are used in forestry: hot fires to promote regeneration and low-intensity fires used for fuel reduction. As J. R. Bartle et al. (1981) note, these are only two of a wide range of possible burning regimes. Even the effects of one type—low-intensity fires—on forest timber yield is variable. It has been found to increase yield in jarrah forests in the short term, to increase yield in the short term but not the medium term in karri and maritime pine forests, and to decrease or not affect yield in spotted gum forests.

Fire is used by foresters not only for bushfire control but also to promote regeneration after logging. In wet forests burning can increase the stocking rate and the growth rates of young trees, but McIlroy (1990) comments that mechanical disturbance seems to promote better regeneration in the Victorian and Tasmanian wet forests than fire. Obviously, however, it is likely also to encourage soil erosion. In dry forests intense burns are not necessary for regeneration, and fires can have two adverse effects. Firstly they may alter the floristics or structure of the understorey; secondly they can stimulate 'green pick' growth, which attracts grazing animals (domestic and native), and these graze on tree seedlings also.

The complexity of fire effects is well shown by McMurtrie and Dewar's (1997) modelling of nitrogen balance in the *Eucalyptus diversicolor* forest of south-western Australia. Short-term pulses of nitrogen are added to the soil by fire, but frequent fires cause nitrogen loss. Nitrogen accumulates in the soil over a long fire-free period as a stand regenerates after harvesting, but is depleted by hazard reduction burns in the later part of the rotation. Trees have their highest stem productivity when young and take advantage of the accumulated nitrogen. Over the whole rotation, nitrogen loss is greatest if fires are frequent and all the trees are harvested. Soil erosion can increase the loss. Of 30 fire regimes modelled, only 10 indicated that the nitrogen balance allowed sustainable yield. However, the researchers noted several caveats—all the growth and nitrogen data were from one stand; there was no measure of nitrogen volatilisation; the spatial variability of the nitrogen fixation measurements was unknown; and the influence of fire regimes on nitrogen-fixing legumes in the forest was not quantified. In this well-studied forest environment, the outcomes of management practices on sustainable yield are reasonably predictable but not fully understood.

Fire management can encourage the spread of woody weeds, yet periodic burning may control the weeds. As fire trails and logging roads were built at the northern end of Barrington Tops plateau in northern New South Wales in the 1950s and 1960s, broom (*Cytisus scoparius*) spread along them. The plant has been in the area since the 1840s, as a garden flower, but had not invaded the bush when vegetation was mapped in the 1930s. Now it has invaded 10 000 ha of open woodland and forest, replacing the native shrubs and shading out the snowgrass and other ground cover. In 1969 part of the plateau was declared national park. The exclusion of seasonal grazing and of fire since then has probably accelerated the spread of the broom, and there are fears that it will move into areas of regenerating commercial forest (Waterhouse 1988).

Dieback of trees in forests and rural areas

In south-western Western Australia most of the land in the high-rainfall zone (with an annual average rainfall of 800–1400 mm) is eucalypt forest that is used for commercial timber production. In the south, woodchip operations are based on clearfelling. Bauxite is mined in the north, and the entire region is the catchment area providing water for most of the State's population. Yet it is a harsh environment for tree growth: summers are dry and fire-prone; the lateritic soils are gravelly near the surface (the source of the bauxite) and clayey and saline at depth. Because of the stored salt, maintaining a healthy tree cover is vital for catchment protection. After the Second World War, as machinery became extensively used in forestry and mining operations, many trees began to show signs of decline (Bartle et al. 1981). Their leaves and branches died, and soon many trees were bare, dead 'stags'. Dieback became an emotive issue. Forestry and mining were blamed for the problem, and the threat to the quality of water supplies from the catchment was taken very seriously.

Trees died because of root attack by pathogenic fungi (mainly *Phytophthora cinnamomi*). The pathogens spread as roads were built into forests for logging and bauxite mining, and they flourished in damp places such as upland depressions on plateaux and along drainage lines. As better controls have been introduced to the main commercial activities, feral pigs and apparently innocuous activities such as recreation and wildflower collection may now be the main causes of the spread of dieback. It affects not only the main timber species, like jarrah, but other trees also (for example, *Banksia grandis*) (Resource Assessment Commission 1992b).

B. Y. Main (1990, p. 405) links dieback with salinisation as ecological disasters and emphasises the secondary ecological effects on fauna:

> The current proliferation of *Phytophthora cinnamomi* in the jarrah forest and other areas is progressing on a scale of comparable magnitude to the regional landscape degradation (salination) of the Wheatbelt, southwest and south coast of Western Australia. One can foresee a similar cataclysmic loss of bird and mammal species dependent on certain species of susceptible flora . . . [and] the disappearance of the unknown invertebrate fauna . . . In the near future those drums, jars and cabinets in museums and herbaria will be the only material evidence of a once rich habitat in south-western Australia pulsing with a diversity of species unparalleled elsewhere outside tropical rainforests.

Dieback is a problem in disturbed forests in other States also. *Phytophthora* was first recorded in Victoria in 1970 in the Brisbane Ranges National Park. It spread along roads and then down creeks. Ten years later over half the areas of susceptible vegetation were infested (Dawson and Weste 1985). B. A. Wilson et al. (1990) reviewed its effects on small mammal populations in a part of Victoria's Otway Ranges, south-west of Geelong, where mining of brown coal, road-testing of agricultural vehicles, clearing, logging, and severe fires in 1983 have degraded the forest. *Phytophthora cinnamomi* infection can kill up to 60% of plant species in the sclerophyll vegetation, and greatly reduce the soil and litter fauna. As a result, the diversity of small mammals in infected areas was as low as in areas burnt in the 1983 wildfires.

The severe dieback that has affected rural Australia (Figure 4.7) has a different cause. Native insects are the cause of this dieback problem, as affected trees are defoliated by chronically high levels of insect predation. Ironically, this

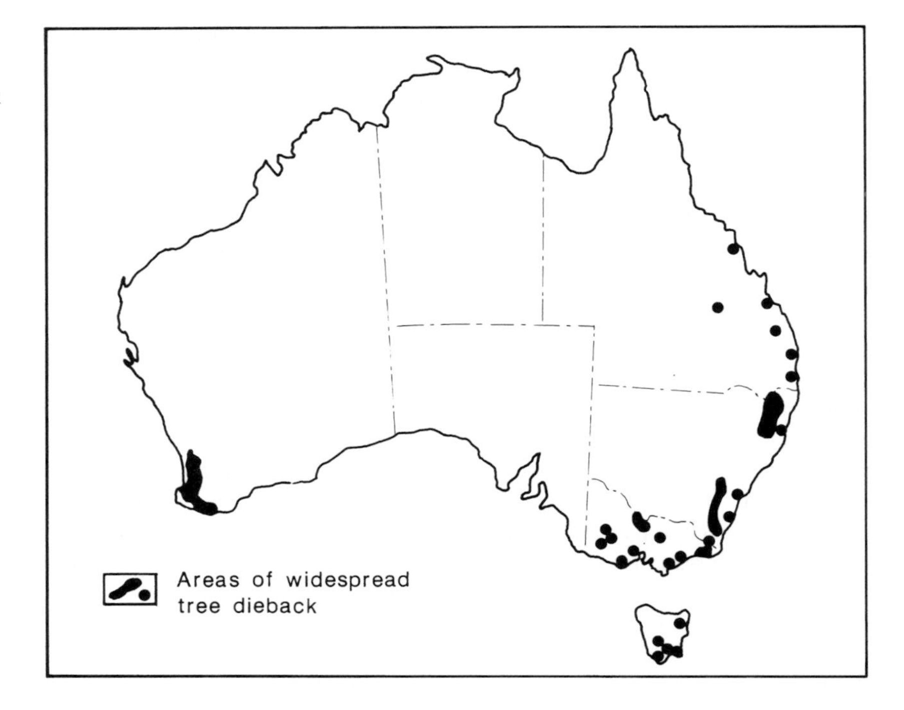

Figure 4.7
Areas affected by widespread dieback of trees. (Source: data from ABS 1992, and Landsberg et al. 1990.)

is the result of better soil and leaf nutrition. A study of isolated stands of trees (mainly yellow box and Blakely's red gum) compared stands that had only pasture grasses on the ground below the trees, with stands that had an understorey of native shrubs and grasses (Landsberg et al. 1990). Those situated amid pasture were more degraded, with more dead or unhealthy trees and fewer birds. Yet they had higher levels of nitrogen in their soil (perhaps from urine from cattle camping below them) and in their leaves. The more nutritious leaves encouraged more insects and hence greater predation on the trees. Other causes may act synergistically; that is, they may work together to accentuate the damage. Root pathogens (such as *Phytophthora* and *Armillaria* species) may also attack the plant; drought can cause stress and open up gaps in the bark, which encourages attack by other insects such as the longicorn beetle; salinity may affect the tree's health. A healthy tree may have three times the leaf weight, one-fifth as many dead branches and five times the root weight of a tree affected by dieback (Heatwole and Lowman 1986).

Dieback in rural areas is a severe problem. It affects the temperate eucalypt woodlands, which have been used for grazing, as well as for timber-getting, and thus were only partly cleared. This partial clearing was thought to be adequate protection against degradation, but serious dieback has negated this view. For this reason it is unfortunate that these woodlands are poorly conserved. Studies in the New England area link dieback to a postwar change in technology and thus land use.

The Walcha area was settled by the 1830s and in 1860 was still largely tree-covered. Now there is less than 1% forest and only 20% woodland cover, and stocking rates have risen from the equivalent of one sheep per ha to as high as eight sheep per ha. This followed the introduction in 1951 of aerial top-dressing with superphosphate, which increased productivity and encouraged graziers to clear more land. Twenty years later, dieback had become a major problem (Heatwole and Lowman 1986).

How is this relevant to the issue of forestry? About half of the forested and wooded lands in New South Wales (seven million of 15 million ha) is used for grazing by sheep and cattle, the grazed areas lying largely outside State forests; in Queensland some claim that nearly all State forests are used for grazing. Grazing has led to changes in the structure and composition of the ground cover and understorey (Resource Assessment Commission 1992c), and dieback in rural areas shows that further indirect and detrimental effects can follow.

Plantations and agroforestry

The National Forest Policy Statement has as one of its goals the expansion of plantations to replace the use of native forests, or at least to minimise further encroachment into them. The aim is to develop plantations on existing agricultural areas, not to clear more public wooded land. In some rural areas softwood plantations are colloquially known as the 'farmer's superannuation'; they use marginal land and promise a retirement income in 20 years' time. Softwoods and other materials such as steel and aluminium are replacing sawn hardwoods in the building industry, and may reduce demand for products from native forests. In South Australia—which has had the longest history of pine plantations and which has little native forest—91% of sawn timber is softwood (Resource Assessment Commission 1992a). However, there is considerable public resistance to softwood plantations. They are inhospitable to native flora and fauna, and are strikingly different to the bushland they often replace.

In 1988 the Australian Conservation Foundation argued for the establishment of large plantations of eucalypts, so that logging of native forests could be phased out, and pulpwood and woodchip production could be replaced by high-value products that depended on a long cutting cycle, like veneers (Mercer 1995). Formby (1991) argued that, in south-eastern New South Wales, a move to softwoods on plantations from woodchips in native forest by 2000 would be feasible and would increase rather than decrease employment. He suggested that the Resource Assessment Commission has underestimated both the strength of community support for forest conservation and the probable competition from overseas eucalypt plantations (in Chile, Brazil, South Africa and Spain). However, while the Resource Assessment Commission (1992a) agreed with the need to develop high-value products, it did not support a suggestion to phase out woodchipping by 2000. At present, woodchipping is still a major use of forests in several states.

As with most issues, industry and conservation groups see the matter differently. To the conservation movement, eucalypt plantations are a substitute for logging in native forests; to the industry, they complement logging. Their advantages include high productivity, and good control of quality and supply. But they also have disadvantages: they are expensive to establish and maintain, and may require share-farming arrangements to provide a large enough area for the plantation to be economically viable. There are environmental costs. Clearing, for planting and during harvesting, can cause erosion; fertilisers and herbicides are used; there are more roads and more frequent entry of machinery; the plantations are monospecific; they do not support the range of shrubs and ground cover plants, or of fauna, found in native forests; they are not managed for very long cutting cycles, so they do not provide the habitats valued by wildlife.

Indeed, it is quite reasonable to regard eucalypts or pines in plantations as agricultural crops. The traditional but artificial distinction between agriculture and forestry is becoming blurred as the new term 'agroforestry' becomes more widely used. Australian pastoralists, particularly in Queensland and the Northern Territory, combine an agricultural use (grazing) with timber production. Although we have already seen (in Chapter 3) that grazing has altered the vegetation below the trees, there are benefits that flow from the practice: destructive fires are less likely; the grazing stock are sheltered by the trees; there is no need to clear completely, so soil and vegetation are protected to some extent. In the southern States the attack on land degradation has concentrated on the replanting of trees, which controls salinity by drawing down the water tables, protects fields from wind erosion by establishing shelter belts, and reduces evaporation from dams. Agroforestry involves the combination of trees, crops and pasture/animals. This can encourage sustainable production and is environmentally preferable to traditional monocultures in agricultural areas (Carne 1993). For the Western Australian wheatbelt, an integrated system called 'alley farming' has been proposed (Lefroy and Hobbs 1992). The aim would be to plant trees to consume the 5–10% of rainfall that is presently unused by crops and is thus contributing to salinisation. Trees would be planted, at close spacing, in belts into which surface runoff would be directed. The belts could also be located to reduce wind erosion and to reunite isolated patches of remnant vegetation. Crops would be grown in the wide and contoured alleys between the shelter belts. This arrangement of trees is best where the trees are using water from the water table, drawing it horizontally. This means the trees are not competing with the more shallowly rooted cereal crops. Also it is compatible with the use of large machinery. Where the trees are in pasture and designed to intercept water before it drains deeply to the water table, a scattered planting is best (CSIRO 1998).

Conclusion

There has been a long history of controversy associated with Australia's forest resources. The need to reserve and protect them was recognised early, but acted on very slowly. Agriculturalists cleared extensively, in the belief that good forests meant good soils, and even that clearing could improve rainfall.

Professional foresters fought a campaign to save the forests and woodlands from agriculture, only to find themselves vilified for destroying natural heritage. Economic development after the Second World War led to increasing demand and to far greater impacts on forested lands. Clearfelling for woodchips to export was the last straw as far as conservationists were concerned, and the fiery controversies about the use of our remaining wooded lands are far from over. Competition for land has been the issue at the centre of the disputes discussed in this chapter: should our remaining forested and wooded areas be used for wood production, agriculture or nature conservation? We will return to the issue of competition later. In the next chapter, we will turn to a land use that occupies little space but has been highly controversial: mining.

Chapter 5

Mining

Introduction

ALTHOUGH it only occupies a small percentage of Australia's land, mining has environmental consequences that impinge on the entire nation. Mining and exploration take place mainly in sparsely populated parts of the country. These are generally the areas in which Aboriginal occupancy has continued since European settlement and are thus the areas on which land-rights disputes are often focused. Ore is processed near the source of minerals in some cases (notably Mt Isa), but more commonly processing occurs in large urban centres; hence problems of air and water pollution as a result of mineral processing affect highly populated areas. Much of our mineral production is exported, and this requires deep-sea ports with sophisticated infrastructure. Port facilities such as these have been developed in most coastal cities, but have also generated urban development in remote coastal areas (notably in north-west Western Australia) (Figure 5.1). Because of the export orientation of the industry, mining is influenced by worldwide environmental concerns, such as the possible impact of fossil fuel use on climate. The industry argues that its continued growth—and the future export income of Australia—are being restricted by political decisions that aim to protect the environment but that can exclude mining development from large areas. Australian involvement in mining in countries once administered by Australia (such as Papua New Guinea and Nauru) raises issues of international relations. Hence, although mining may occupy directly less than 0.02% of the land surface and related infrastructure only about 0.4% (Lambert and Perkin 1998), it is difficult to agree with Brooks (1988) when he comments that the effects of mining are localised.

'Transformed by metals': The development of the industry

It was the discovery of gold in New South Wales in 1851—and soon after in Victoria—that began the first mining rush in Australia. Today it is our second most valuable export ($6.2 billion in 1997–98), surpassed only by black coal ($9.2 billion) and surpassing wool ($4.1 billion), iron ore ($3.9 billion), and wheat ($3.8 billion) (ABARE 1999). In the 1850s, it transformed Australian society:

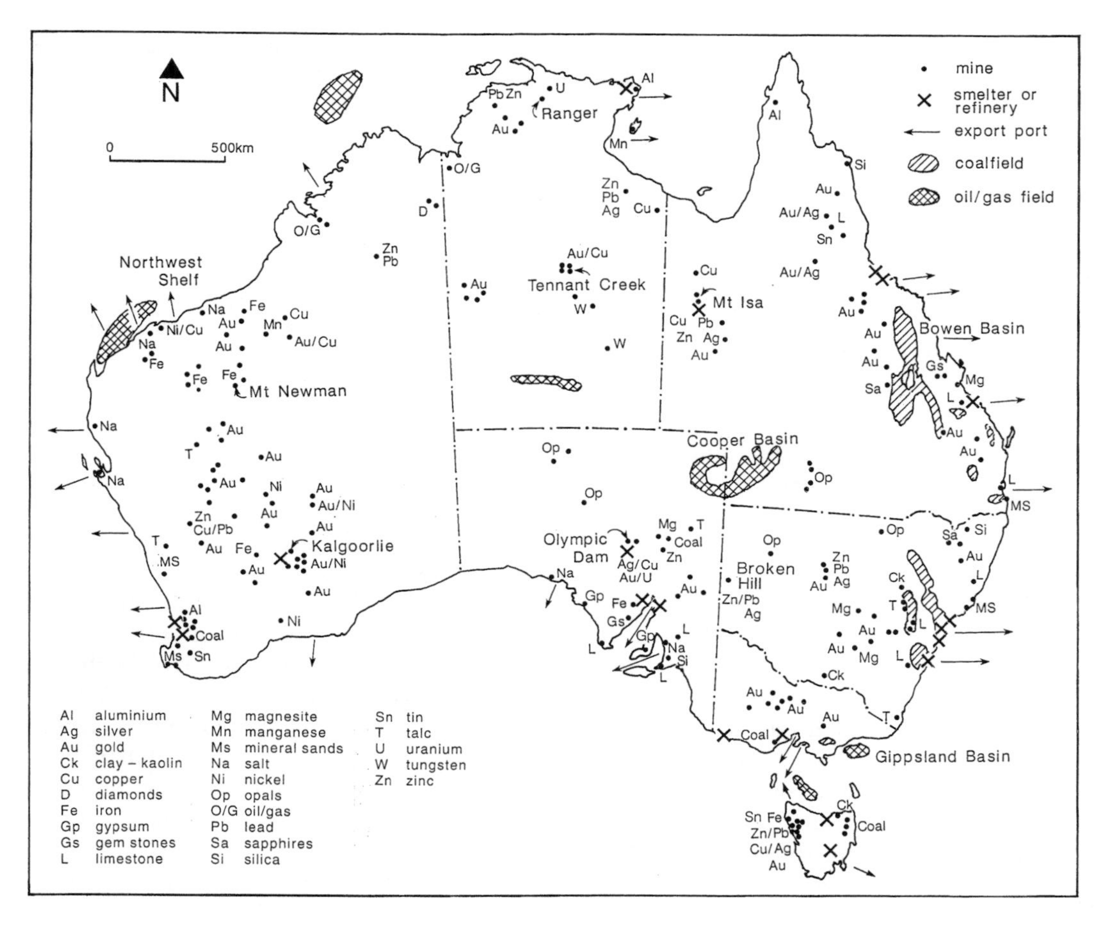

Figure 5.1 Locations of mines, and of mineral processing and export. (Source: redrawn from Australian Mining Industry Council 1992, and Lambert and Perkin 1998.)

> The rise of the goldfields had thrown cities and farms and sheep runs out of balance, depriving them of men, but ultimately these pursuits were so stimulated by gold's wealth and immigration that at the end of the 1850s they were depriving the goldmines of labour and capital and accelerating their decline . . . Gold had stimulated such an expansion of Australia's economic fabric that the economy became strong enough to cushion mining's fall. Possibly no other country in the world had been so quickly transformed by metals; the normal growth and achievement of several decades were crammed into one.
>
> Blainey 1963, p. 61

This transformation had a profound influence on the environment, both locally on the gold-fields themselves (see Figure 1.4, p. 5) and more widely in pastoral and cropping areas. A vivid picture of the scene at Ballarat is painted in Henry Handel Richardson's novel, *Australia Felix*:

> No patch of green offered rest to the eye; not a tree, hardly a stunted bush had been left standing, either on the bottom of the vast shallow basin itself, or on the several hillocks that dotted it and formed its sides. Even the most prominent of these . . .

> had been stripped of its dense timber, feverishly disembowelled, and was now become a bald protuberance strewn with gravel and clay . . . Seen nearer at hand, the dun-coloured desert resolved itself into uncountable pimpling clay and mud-heaps . . . Water meandered over this mud, or carved its soft way in channels; it lay about in puddles, thick and dark as coffee grounds; it filled shallow abandoned holes to the brim.
>
> Richardson 1978 [1917], pp. 5, 6

Clearing of the trees, disruption of drainage systems, excessive sedimentation into streams, erosion of soils, careless placement of sterile spoil, and destruction of alluvial flats commonly accompanied gold-digging.

The search for gold was not confined to the surface. At the Coolgardie and Kalgoorlie finds in Western Australia, which developed in the 1890s, miners moved underground to take gold from the rock itself (Blainey 1963). Today the Mount Charlotte mine at Kalgoorlie has one shaft that is 1180 m deep (Australian Mining Industry Council 1992). At Ballarat gold was mined from shafts—up to 600 m deep—that tapped into the 'deep leads'. These were old river sediments, infilling channels that had been covered by basalts that poured out over the land during the Pliocene period. Their gold was alluvial, but was now buried deep in the landscape. Similar concentrations of other alluvial minerals were mined in deep leads elsewhere, such as the tin fields of Herberton in Queensland (Figure 5.1), and Ringarooma in Tasmania (Figure 5.2). Yet even after extensive exploration and mining, the courses of the buried rivers remain disputed and gold exploration in Ballarat continues with the question of long-term drainage development unresolved. Interestingly the two views on where best to seek the buried alluvium is a practical outworking of the two major interpretations of the long-term evolution of the landscape in south-eastern Australia—stability of drainage networks, or divide migration and stream capture (Bishop and Li 1997, Young and McDougall 1993).

The fields around Kalgoorlie serve to focus our attention on some important social consequences of gold mining. These fields were important enough to demand and obtain major resources from the populated and well-watered coastal region. The railway came in 1896, and water followed in 1903 via a long pipeline from the Darling Range. These were early instances of a now common phenomenon in Australia: the large town dependent on mining and set in inhospitable arid areas at great distance from other major urban centres. Interestingly, the populations of the Kalgoorlie fields did not include a large Chinese contingent, unlike the populations of all the earlier eastern and northern gold-fields. Chinese immigrants were not welcomed anywhere, but had outnumbered miners of European origin in New South Wales by the late 1850s, and in Queensland soon after. In the Northern Territory they were exploited at Pine Creek as cheap labour in the 1870s. By the 1880s, when fields in Western Australia opened, all colonial governments had united against them, and they could not move into the new fields (Blainey 1963).

Important though gold was, it was not the focus for the first major mining activity in Australia. In the 1840s copper was mined at Kapunda and Burra, and this made South Australia the most prosperous of the colonies by the end of

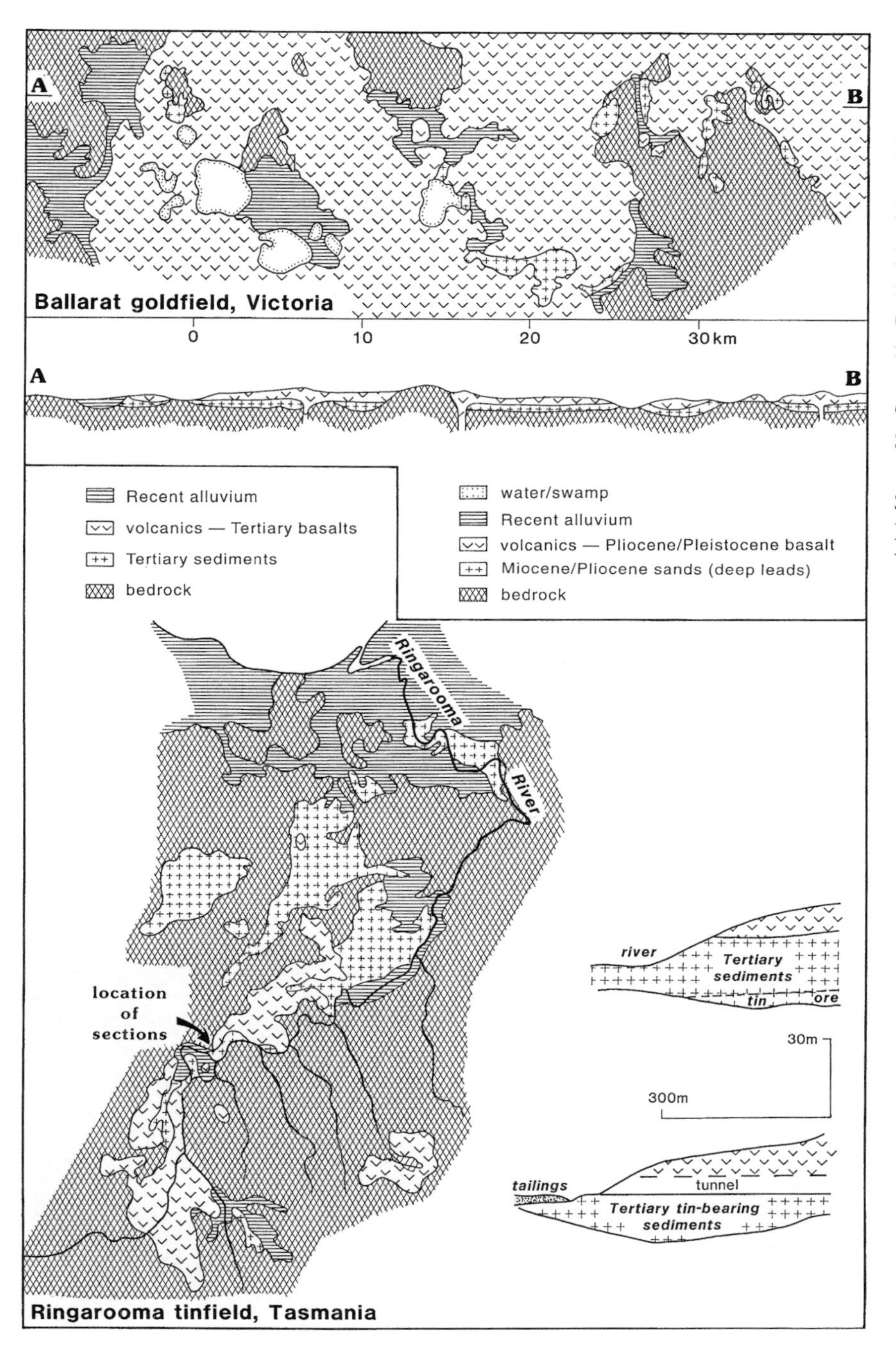

Figure 5.2
Tertiary sediments (ore-bearing deep leads) underlying basalts in the Ballarat gold-field, Victoria, and the Ringarooma tin-field, Tasmania. (Source: redrawn from the Ballarat 1:250 000 Geological Sheet SJ54-8, 1st edn, 1973, Geological Survey of Victoria, Melbourne, and Yim 1991.)

that decade. After a hiatus due to a loss of labour to the gold-fields, other mines opened at Kadina and Wallaroo, and the 36 furnaces there consumed 10% of the coal mined from Newcastle in the 1860s. Backloading of the coal ships with low grade ore led to the development of smelting facilities at Newcastle (Blainey 1963), and this exemplifies again the way in which the environmental impacts of mining rapidly extended well beyond the orefield.

As the gold-fields opened up in the West, so did the mining of silver-lead deposits at Broken Hill, where the smelters started up in 1886. This provided the impetus for exploration in western Tasmania, leading to the developments at Zeehan in 1891 and the Mt Lyell deposit at Queenstown in 1896. The environmental consequences of mining and smelting at Queenstown were so dramatic that the barren landscape remains a tourist attraction to this day (Figure 5.3). The bare and eroded hills stand in stark contrast to the dark green and dense tree cover on unaffected areas upwind of the town. Sulphurous and acidic fumes with high heavy-metal content were discharged by large wood-fired smelters (Figure 5.4). Cutting for fuel for the smelters, and damage by the pollutants, completely removed the tree cover from the valley and hills downwind. The shallow, organically rich soil that supports the forests in this region was quickly washed away, exposing a subsoil that was often no more than weathered bedrock and had high levels of heavy metals. Vegetation could not re-establish itself on the steep slopes, and the valley is often described as a lunar landscape. It is certainly an extreme example of the severity of the environmental damage that can be caused by mining, and of the persistence of that damage.

An assessment of mineral resources shortly after the Second World War noted that Mt Isa (lead, silver, zinc and copper) was the only major new field, and that the search for oil and gas had been disappointing (Edwards 1949). This was soon to change. New technology led to a demand for new minerals: uranium, mineral sands and bauxite. Increasing demand encouraged exploration and major developments in the 1950s and 1960s: bauxite at Weipa

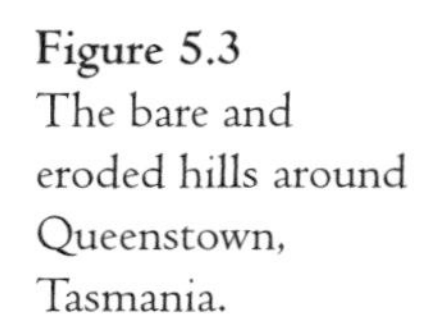

Figure 5.3
The bare and eroded hills around Queenstown, Tasmania.

Figure 5.4 Sulphurous smoke belching from the wood-fired smelters at Queenstown, c. 1890. (Source: the Zeehan Mineral Museum.)

(Queensland) and the Darling Range (Western Australia), iron ore in the Hamersley Range (Western Australia), uranium at Mary Kathleen (Queensland), manganese at Groote Eylandt (Northern Territory) and nickel at Kambalda (Western Australia). A decade later an industrial boom in Japan, and concern over oil supplies worldwide, triggered rapid growth in exports of coal and iron ore (Box 8).

Box 8
Mining in the Pilbara

The Hamersley Basin in the Pilbara region of Western Australia contains about 90% of Australia's vast identified resources of iron ore. The ore occurs primarily as banded iron formations. This type of sedimentary rock was deposited during the Proterozoic era (beginning 2500 million years ago), as photosynthesis by newly evolved cyanobacteria and algae released oxygen into the earth's atmosphere, and vast quantities of iron, previously in solution in the world's oceans, were oxidised and precipitated. In the Pilbara, metamorphic processes have further enriched the iron content of the rocks. Over geological time some strata have been eroded, and debris from these strata has been redeposited in palaeochannels and recemented by more iron; these pisolitic deposits are also mined. Prior to the Second World War there was concern that Australia's iron ore reserves were too small to support a domestic and an export market, so in 1938 an embargo was placed on exports. When this was lifted in 1960 a spectacular rise in both exports and estimates of resources took place, as development and exploration boomed in response to strong demand from Japan.

The industry in this region is still growing, with new mines being opened (usually sharing existing infrastructure) and the capacity of some older mines being expanded. Very little new iron ore development is proposed outside the Pilbara, and the largest resource development in Australia's history—the North West Shelf project—is located off the coast of the region (Ecologically Sustainable Development Working Groups 1991c) (Figure 5.5). This project supplies gas by pipeline to the State Energy Commission of Western Australia, and since 1989 has shipped liquefied natural gas (LNG) to large Japanese power and gas companies. Australia claimed control of the continental shelf under customary rights of international law in 1953, and this allowed the

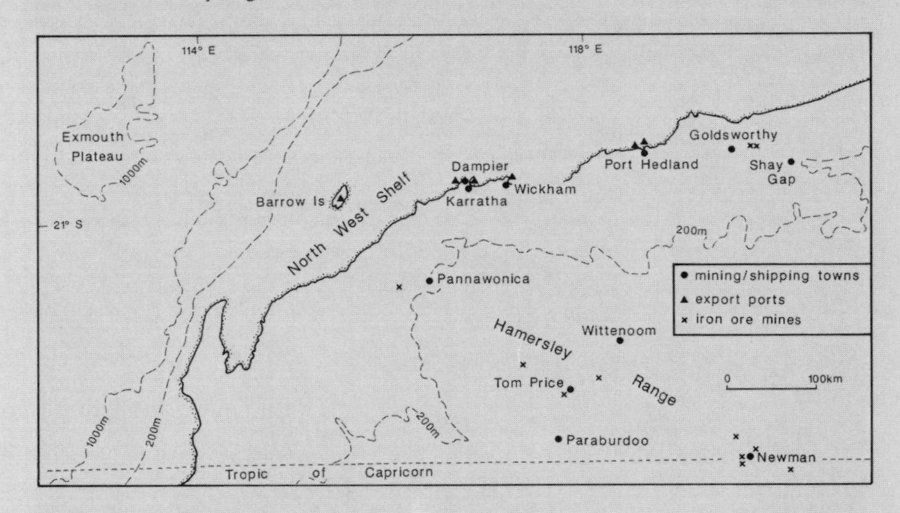

Figure 5.5
Mining developments in the Pilbara region, Western Australia.

Figure 5.6 Mt Tom Price mine, Hamersley Range. The vastness of the mine is shown by the scale of the long ore train in the background. (Source: photo by B. G. Jones.)

development of both the North West Shelf, and the Bass Strait oil and gas fields.

Iron ore projects led to the development of ten new towns, four new railways, three new deep-water ports and a tenfold population increase in the Pilbara between 1962 and 1975. Pastoralism—established since the 1860s—had previously been the predominant land use. Aboriginal people had worked on pastoral properties, but under contracts that were so restrictive that in 1946 they organised the first major Aboriginal strike and set up small, independent companies for pearl shelling, contracting and some mining. They were unable to participate in the new large-scale mining ventures which recruited workers from the capital cities so that labour would not be drawn away from regional rural activities. As R. Howitt (1989) comments, 'In comparison to the relative wealth of the new towns, the poverty and powerlessness of most Aborigines was reinforced'. Ironically, problems increased when national legislation to give Aborigines greater equality was passed in the late 1960s. This gave them legal access to alcohol (in 1967) and required that they be paid wages equal to those of White rural workers (in 1968). Both changes encouraged the drift of Aborigines from pastoral properties into town camps, where there was little opportunity for work. The effects of the economic restructuring of the region, allied with the changing social and legislative attitudes towards Aborigines, had profoundly detrimental, although unintended, effects on the Aboriginal people of the Pilbara. In Western Australia less than 1% of the mining workforce is Aboriginal (and over half of those are employed on a casual basis). This figure is significantly lower than the percentage of Aborigines in the population as a whole, particularly in remote areas.

The Pilbara was the source of Australia's most controversial mineral: blue asbestos (Plate I). It was mined at Wittenoom between 1943 and 1966, and promoted as a 'magic mineral' because it resisted fire, acid and salt corrosion. It was fibrous and could be woven into filtering or insulating material; it added strength to building materials and was an important component of the 'fibro' of which many Australian houses were built. But it was a deadly mineral. There is no safe level of exposure to blue asbestos, and the result of exposure is severe lung disease (asbestosis) and lung cancer (mesothelioma). The links between blue asbestos and these diseases have been known since the 1930s. However, it was only in the 1980s, long after the mine at Wittenoom closed, that the legal battles against the mining company for compensation for affected workers prompted action to protect people from exposure (such as the careful removal of asbestos-containing building materials from homes and public buildings). Indeed it was only in March 1994 that the first compensation for passive exposure (rather than for exposure in asbestos mining or processing) was awarded in Australia; a clerical worker in a local council was exposed to asbestos that had been sprayed onto the ceiling for acoustic control, but not sealed, and the worker developed mesothelioma.

Data from Anon. 1993, Bambrick 1979, Ecologically Sustainable Development Working Groups 1991, Hills 1989, Morrison 1988, North West Shelf Joint Venturers 1993

Mineral resources and sustainability

Australia is a major global player, with the world's largest economic demonstrated resources of bauxite, lead, mineral sands, silver, tantalum, uranium and zinc. It is among the top six with respect to resources of coal (brown and black), cobalt, copper, gold, iron ore, nickel, industrial diamond, lithium and rare-earth oxides (Lambert and Perkin 1998). Global questions of resource availability and also of ecologically sustainable development therefore are highly significant issues for mining in Australia.

There was concern in the 1970s that non-renewable resources would limit economic growth. More recently, the focus has changed to concern that 'limits are being set by the environmental impacts of increasing human population and its increasing consumption of resources' (Rutland 1997, p. 13). Over the past few decades, the intensity of use of minerals (measured in kg/$GNP) has declined due to recycling, greater efficiency of use, substitution with non-metals such as plastics, and higher value-adding of finished products. However, the total consumption has only stabilised or declined in developed countries. In the newly industrialising countries such as Korea, consumption has increased —providing, of course, export markets for Australian minerals. Thus, worldwide consumption is increasing as population increases and as countries strive for economic development (Rutland 1997).

There are two aspects to sustainable use of mineral resources, the first of which is continuing availability. Lambert and Perkin (1998, p. 12) comment:

> The minerals industry now generally accepts that there should be some special areas where resource development should not occur. But continuing the trend to set aside large other areas of land for conservation, to the exclusion of other activities, is clearly not sustainable in the long term ... It is not possible to say with any authority that a given area ... is barren of any economic mineral resources.

Two examples illustrate their concerns. About 30% of Australia's rutile and zircon resources cannot be mined because they are reserved in national parks along the Queensland and New South Wales coasts. Reassessment of the commercial viability of a huge phosphate reserve near Duchess in northern Queensland has moved Australia from a position of almost total dependence on imports, to a resource ranking of 7th in the world.

The second aspect concerns the environmental impacts of mineral extraction, processing and use. Most international attention, for example via Agenda 21, has focused on a few issues, especially rising CO_2 levels and climate, and management of wastes. Other matters are relevant both globally and within Australia. Much attention has been paid to metallic ores and the impacts of their mining and processing (such as sulphurous emissions from smelters, or acid drainage from waste dumps. Yet in major industrialised countries, the greatest tonnage (per person) of minerals used is of construction materials, then fuel minerals; metals make up a relatively small percentage of the total (Rutland 1997). Near Sydney, the need for sand for construction purposes was met initially mainly from coastal dunes, then driven inland to river terraces and river beds as coastal erosion after major storms in the mid-1970s triggered protests against dune removal. This move has had significant impacts—loss of good agricultural land as river terraces were mined, stream erosion after dredging, and localised stream turbidity. Yet the demand continues and recurrent proposals to take sand from the nearshore continental shelf trigger more protests than does the dredging of rivers.

Environmental concerns have led to a revised view of waste management. This should not be seen simply as actions needed at the end of a process, but as a re-evaluation of all the process. For example, water pollution by saline runoff from coal mines is reduced by re-use of the waste water for dust suppression on site. The emission of methane from underground collieries and from landfill sites has been reduced by installation of power plants which use the methane to fire gas turbines. The gas generators using methane from collieries near Appin, New South Wales, are claimed to reduce Australia's output of greenhouse gases by 0.5%. Nevertheless, the impacts associated with mining continue to be appreciable.

The impacts of mining

The worst examples of environmental degradation caused by mining are the older areas such as Queenstown in Tasmania, Captains Flat in New South Wales and Rum Jungle in the Northern Territory. At Captains Flat—on the Molonglo River, 50 km upstream of Canberra—zinc, copper, gold and pyrite were mined between 1874 and 1962. Tailings dumps—spread over 15 ha, with high metal concentrations and high acidity—remained uncovered. They collapsed and washed into the river, polluting the water and killing pastures onto which sediment was deposited downstream. Rehabilitation works since 1979 have controlled the problem, but at great public expense. At Rum Jungle the contaminants also spread downstream, but in this case there was added concern due to the radioactive nature of the waste. Again rehabilitation has been at public expense (Bell 1986). Since the 1970s, as in other aspects of environmental management, attitudes and practices in mining have changed markedly. Even so, critics argue that 'best available technology is not applied consistently throughout the industry', especially in operations established more than 20 years ago or distant from major population centres (Ecologically Sustainable Development Working Groups 1991c, p. 25). Nevertheless, the mining industry is placing increasing emphasis on environmental auditing and responsibility. In 1996 the Australian Minerals Industry's Code of Environmental Management was launched and its 41 signatory companies produce about 80% of Australia's minerals. While acknowledging this progress and that of the petroleum industry's sister code, some non-government organisations see the code as too limited. It does not provide for an independent monitoring body or ombudsman, and does not address issues of human rights and land acquisition when companies are operating outside Australia (Australian Minerals and Energy Environment Foundation 1999). An added dimension recently has been the pressure exerted by shareholder groups. Via action at company general meetings, some groups have sought to influence the managements of mining and other resource companies to alter policy to a more environmentally responsible path.

A measure of the impact of past practices is given in A. D. Knighton's (1987) study of the alluvial tin mining industry of north-eastern Tasmania. Tin mining began along the Ringarooma River in 1875. Dams and water races were built to provide high pressure for sluicing operations. Jets of water broke up Tertiary alluvial deposits and directed the sediment into sluicing boxes. The heavy tin ore (cassiterite) dropped out in the boxes, and then the waste sediment was discharged to the nearest stream. Later dredging was introduced, but again the product was processed on land, and the unwanted sediment was put back in the stream. More than 70% of the ore was produced by mines within 1 km of the Ringarooma River, so the impact was direct and severe. Bridge records show a slow aggradation near the mines of 4–6 m between 1930 and 1970, and sediment is still moving down the river. About 30 km downstream the estuary has silted up and a wharf that once served ocean-going ships now stands on dry land. Knighton estimates that 40 million m^3 was added to the river by the time mining declined in the 1980s. Similar disruptions were caused in many other places, and ironically some now are tourist attractions (see Plate A).

Prior to the 1970s, few mining companies other than mineral sand and bauxite operators saved topsoil for rehabilitation works. Yet soil management is central to reducing the impact of mining (Bell 1986). Spoil from mining, or surfaces exposed by mining, are often characterised by:

- extreme acidity (pH 2.5–4.0) or alkalinity (pH 9–11)
- toxic concentrations of metals
- excessive salinity (EC_e 4–20 mS/cm) and sodicity (ESP 15–60)
- crusting, poor permeability and/or low water-holding capacity
- dearth of micro-organisms and soil fauna
- low organic matter content and lack of seed store.

The spoil may be weathered bedrock or subsoil now brought to the surface (during open-cut operations), sediment that has been reworked (as in alluvial mining), or waste rock from which minerals have been separated (for example, by flotation separation of coal or base metal ores). In all these cases it is unlikely that the spoil will be a good medium for plant growth. Similarly, mining can expose unsuitable material. For example, at Weipa on Cape York the B horizon of soils in the areas mined for bauxite has a high capacity to immobilise phosphorus, so that regeneration on exposed or respread material from this horizon is poor. Soil horizons are therefore stripped in tandem and replaced in the correct vertical order (with material from the A horizon above that from the B horizon) to encourage regeneration without the need for excessive additions of fertiliser (Bell 1986).

Between Perth and Bunbury in Western Australia bauxite deposits in the jarrah forest have been mined since 1963 (Grant and Koch 1997, Koch and Ward 1994). The pits are shallow (averaging 4 m in depth) and 1–10 ha in extent, and 450 ha are mined and rehabilitated annually. Before mining, topsoil is stripped. The upper 5 cm is taken first, and then the remaining A horizon (usually less than 35 cm). These layers may be directly replaced in a mined pit or stored for replacement later. After mining the topsoil is replaced in correct order, a mix of seeds is broadcast, and fertiliser is applied. The aim is to re-establish the pre-mining vegetation. The broadcast seed includes native legumes, which grow rapidly, and build up the litter and nitrogen content of the soils. Adequate nitrogen levels (from the legume activity and from fertiliser) do not affect plant density, but do increase plant cover and species richness in revegetating areas. New rehabilitation is dominated by the fast-growing annuals and biennials that are prolific seeders. In this they are similar to areas affected by fires. Mature forest is dominated by slower-growing, longer-living plants that resprout after fire and do not produce large quantities of seed. Thus, the topsoil does not contain many seeds from these plants, but there is significant recruitment from surrounding forest. Overall, only 23% of the non-weed regeneration came from broadcast seed, and the remaining 77% from topsoil and recruitment. Weeds are not common. The species that are missing from regenerated areas are the understorey plants, which are difficult to establish: orchids, rushes, and sedges. However, they may reappear over time. After one year, the re-established vegetation may have a similarity index of only about 25%, relative to mature forest. After ten years that index rises to over 50%, a figure similar to the index for forest-to-forest comparisons. The full pattern will not be clear for some years yet. Initially revegetation was with pines. After 1976, native understorey seed was used but, because of dieback problems, some *Phytophthora*-resistant non-endemic eucalypt seed was sown. Since 1986 jarrah and marri overstorey has been sown. Recently also, seed has been treated with heat and with smoke to promote germination.

In other countries acidification of soil and drainage due to oxidation of iron sulphide (pyrite) in coal is a major problem. The oxidation process is the same as for acid sulphate soils (see Box 5), and the drainage is polluted by acidity and by high concentrations of iron and other metals. In Australia the coals have relatively low sulphur contents, but the fields in Western Australia, the Hunter Valley of New South Wales and the Bowen Basin of Queensland have some high-sulphur seams. Obviously, the same problems are found in areas where metal sulphide ores are found, and again Queenstown and Captains Flat provide extreme examples. The sulphide is also oxidised as the ores are smelted, leading to the air pollution that denuded the Queenstown hills and that remains a matter for concern near smelters today. The impact of air pollution from smelters is due to both sulphur oxide emissions (which are linked to asthma incidence, and may cause acid rain) and the release of metals, such as lead, in the fumes (Box 9). Where coal is used to generate electricity, the main concerns are sulphur oxides, nitrogen oxides (which contribute to photochemical smog) and particulate emissions (which reduce visibility) (Jakeman and Simpson 1987). The brown coal resources of the Latrobe Valley in Victoria and the black coal deposits of the Hunter Valley in New South Wales are the main areas of concern.

Since 1973 soil conservation and rehabilitation works have been mandatory in coal mines in New South Wales (Hannan 1984). The change reflected not only growing concern for the environment generally but also a change in mining technique. Most of the early Hunter fields had been worked by underground mining, but new technology and competition with the Bowen Basin fields led to development of open-cut or strip mining. This replaced former pastoral activity, particularly in the upper Hunter Valley, near Muswellbrook (see Figure 2.5, p. 21). Clearly, this was a far more disruptive and obvious method of mining, and conflict between mining and agricultural interests resulted (see Chapter 2). Mining involves the removal of overburden from a strip of land, placement of the debris in long heaps beside the pit and mining of the seam at depth. It advances across the country in a long front, leaving behind it mounds of debris that are progressively reshaped and rehabilitated (Figure 5.7).

The rehabilitation works address a range of issues (Bell 1986, Hannan 1984):

- The mined overburden swells and occupies a greater volume than before. Slopes on the mounds, even after compaction and reshaping, are higher and

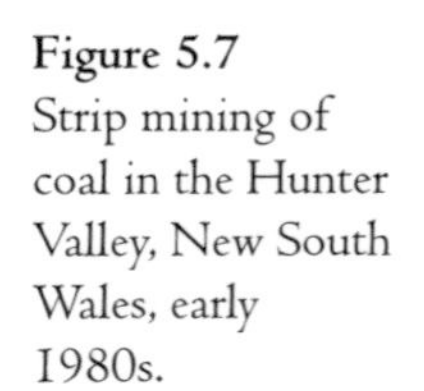

Figure 5.7 Strip mining of coal in the Hunter Valley, New South Wales, early 1980s.

steeper than slopes in the undisturbed landscape. Denser drainage networks need to be constructed to prevent erosion on these new slopes.

- Weeds and vermin can invade the disturbed areas. Windrows of cleared trees need to be burnt to stop rabbits and other pests sheltering in them. Disturbed sites need to be cultivated to eliminate weeds. For example, in the Hunter Valley a weed, *Galenia secunda*, is not found in native undisturbed pasture but rapidly invades disturbed ground because it has low nutrient requirements and a deep rooting system that allows it to withstand drought.
- Spoil that is acidic, saline or sodic needs to be buried and not left near the surface. Even during high rainfalls, seepage into the spoil heaps only penetrates 4–5 m so material can be buried at this depth (provided this is well above the water table, so that groundwater is also protected).
- Land is usually returned to pastoral use, and it can be argued that, at least in the short term, productivity may be higher than previously. One significant social change is that, although the land use might have been reinstated, the ownership of the land is likely to have changed, possibly from a family company to a major national or international company.
- Control of drainage is necessary to prevent release of saline groundwater encountered at depth in the mines. In the Hunter Valley there are shallow, fresh aquifers associated with Quaternary alluvium along the valleys, but aquifers in the Permian coal-bearing beds are saline. Saline water is retained and reused on site (for dust suppression, irrigation or coal processing), or stored and discharged under licence to the rivers during high-flow periods.

The aquifers, and the alluvium with which they are associated, are protected from mining by legislation, even though there are large reserves of coal under prime agricultural land in the Hunter Valley, in the Liverpool Plains further north, and in south-eastern Queensland.

Box 9
Lead in the environment

There is a worldwide trend to reduce the use of lead and restrict emissions of lead into the environment. For example, many countries, including Australia, have introduced legislation and price incentives to reduce the use of petrol that contains lead as a catalysing agent, and lead-free paints are now common. Lead can cause neurological and behavioural disorders, particularly in very young children, and medical authorities consider that there is no safe minimum level of exposure. Hence, both the USA and OECD countries are promoting strategies that will reduce lead use. These have serious implications for the Australian mining industry. Australia is the world's largest primary producer of lead. As the proposed strategies will affect lead prices, the viability of many of our mines may be affected. This is because lead is an important product of many mines that also produce zinc, copper and silver, and the overall viability of these mines is related to prices for all these metals.

In Australia concern about the health of people exposed to lead was focused initially on the town of Port Pirie in South Australia. The lead smelter there has been operating since 1889, and for many years there was little effort to contain the contaminated dusts and fumes from the plant. In 1925 a Royal Commission found that there had been 383 cases of lead poisoning (plumbism) among workers since 1910, and better occupational health procedures were introduced. However fugitive lead emissions remained a problem until the early 1980s and increased the lead content of soils in the town. The blood lead levels of people living in the town correlated well with the lead contents of the soils around their homes. In young children, blood lead levels were higher than in adults; the levels peaked at about two years of age, apparently because small children ingested lead-contaminated dirt and dust when playing. Studies showed a statistically significant relationship between increased blood lead levels and poorer than average mental development. Lead levels still exceed National Health and Medical Research Council (NHMRC) guidelines near the works and wharf stockpiles.

Correlation between soil lead and blood lead levels were demonstrated in several studies carried out near smelters in New South Wales. In Wollongong 8.4% of preschool-aged children living near the smelter had blood lead levels above the level of concern set by the NHMRC (25 μg/dL at the time of the study in 1990). That level of concern was reduced to 15 μg/dL in 1994, and 54% of the children tested had blood lead levels above that figure. In a 1992 study at Boolaroo, near Newcastle, 6% of children had levels above 25 mg/dL and 49% above 15 μg/dL. At Broken Hill, with a long history of mining and smelting of lead ores, 19.8% of children had levels above 25 μg/dL. The current NHMRC goal is to have 90% of all children with blood lead levels below

10 μg/dL. Not surprisingly both Commonwealth and State health and environmental authorities are developing strategies to reduce community exposure to lead, both in the major cities and near lead processing plants.

Data from Baghurst et al. 1987, Body et al. 1988, Brown et al. 1991, Department of Environment Sport and Territories 1996, Environment Protection Authority (NSW) 1993, Young et al. 1992

Subsidence and alteration to groundwater flow

While the environmental effects of strip mining of coal are obvious, underground extraction also causes significant impacts. Surface facilities and transport infrastructure are required; and the movement of coal by road, and through urban areas to ports, power stations or steel mills, has been a contentious issue in all New South Wales coalfields. The impacts of subsidence into the underground workings were the focus of controversies, particularly from the mid-1970s to the early 1980s, when the coal industry was expanding.

As coal is extracted, the roof of the worked area is allowed to collapse, to reduce stresses on the new working areas and thus to increase mine safety. The ground surface above the worked area, and for some distance around it, subsides as the roof collapses. How much subsidence occurs is determined by several factors:

- the depth of cover (that is, the depth to the coal seam from the surface)
- the stratigraphy (for example, whether there are thick beds of competent sandstone to bridge the gap and support the surface
- the presence of faults or dykes, as these may distort the subsidence profile
- the number of seams mined and their thickness
- the angle of draw, which measures how far subsidence extends beyond the worked area.

In the coalfields around Sydney and in the Hunter Valley, the coal leases are adjacent to national parks; in the southern coalfield of New South Wales, west of Wollongong, the coal leases underlie the metropolitan water catchment for Sydney. In the mid-1970s the Water Board argued that coal should not be extracted from below stored waters, dam walls or any area within a 35° angle of draw of these. The coal companies protested that this would unnecessarily sterilise coal reserves and reduce the viability of the mines. An inquiry (Reynolds 1977) rejected the suggestion that catastrophic leakage from the reservoirs could occur, but accepted that changes in the pattern of groundwater could cause some loss of water. It recommended mining, but with restrictions and further investigations.

In the southern coalfield, the average depth of cover to the main seam—the Bulli Seam of the Illawarra Coal Measures—is 190–310 m. Above an area where two seams were mined and a dyke distorted the profile, the ground subsided a maximum of 2.4 m over a distance about 1.6 km long and 1 km wide. Cracks up to 600 mm wide and 9 m deep formed, running in the same directions as cliffs and ridges (Kapp 1980). At Appin, where the cover was deeper (500 m) and comprised largely of strong sandstones, subsidence barely affected two historic churches; although they dropped by 110 mm and 530 mm, only minor cracking of the structures resulted (Kapp 1982). The F6 expressway from Wollongong to Sydney was less fortunate. Extraction of coal at a depth of 400 m led to subsidence of 1.4 m, and the kerbing, gutters and road pavement were damaged (Castle et al. 1983). However, it is the collapse of cliffs, triggered by subsidence and related changes in stresses, that has fuelled the resistance to coal mining in national parks. Above undermined areas, stress is concentrated in valley floors and these can be disrupted by cracks and displaced blocks, allowing flow to go below the creek bed. This is obvious particularly during low flow, and when the flow reappears downstream it may be low in oxygen and high in precipitated iron, giving conditions unfavourable for aquatic life. A study in Dharawal State Recreation Area, in the headwaters of the Georges River, showed cracking and disruption in the sandstone beds of steep-sided creeks to be common above the areas undermined by longwall coal extraction, and absent outside these areas. The most controversial impact of subsidence in the coalfields west of Wollongong has been the leakage of methane into the bed of the Cataract River. In 1997 this caused the death of surrounding vegetation and of aquatic life, and forced the erection of signs warning against smoking nearby. The low flow in the river was due to drought and to water storage upstream for the Sydney supply, considered to be contributory factors to the effects of subsidence (Searle 1997).

In 1980 a zoning scheme to allow underground coal mining in national parks in New South Wales was proposed (Department of Environment and Planning 1980). It recognised the potential for problems, such as landslides, cliff collapse, stream capture, subterranean stream drainage, rehabilitation of tracks and workings, and water pollution. However, its proposed scheme was clearly unrealistic, and completely unrelated to the real distribution of coal or the practical necessities of mine location within any area. For example, it proposed helicopter access to mining-related sites within sensitive environments, a suggestion hardly compatible with the aim of preserving a sense of naturalness in the remoter parts of a park. Historically, major cliff collapses in the region have been associated with areas where full extraction of coal has taken place, particularly in the Blue Mountains, where the depth of cover may be shallow (Young and Young 1988). The former tourist vantage point of Hassans Walls, near Lithgow, now has restricted access because of the severe and continuing collapses along its cliffs. At North Nattai, beside Lake Burragorang, a rockfall 210 m high and 775 m long deposited 30 million tonnes of debris over a distance of 1 km between 1965 and the mid-1980s. Coal was mined from below and in front of the cliff in the 1950s (Cunningham 1988). Detailed analysis of rock falls in an area of longwall coal extraction, at cover depths of 43–212 m showed that mining subsidence generated new cracks along cliff lines, and that cliffs that were not undermined did not fail. Sections of cliff were compressed from the sides and pushed outwards from the plateau as the mining occurred below them. Only 16% of the cliffs that might have failed actually did so (Kay and Carter 1992), but this percentage is too high in the opinion of some conservationists who vigorously but unsuccessfully opposed the extensions of some coal leases in the district on the grounds that subsidence would damage the 'pagoda' formations on the Newnes Plateau (Figure 5.8).

Plate A
A hillside excavated by sluicing to mine a deep lead, near Kiandra, New South Wales. The site is developed as a tourism feature.

Plate B
A party of tourists watching sea lions, and the boardwalk they have used to gain access to the beach, are dwarfed by the dunes and the aeolianite cliffs of southern Kangaroo Island.

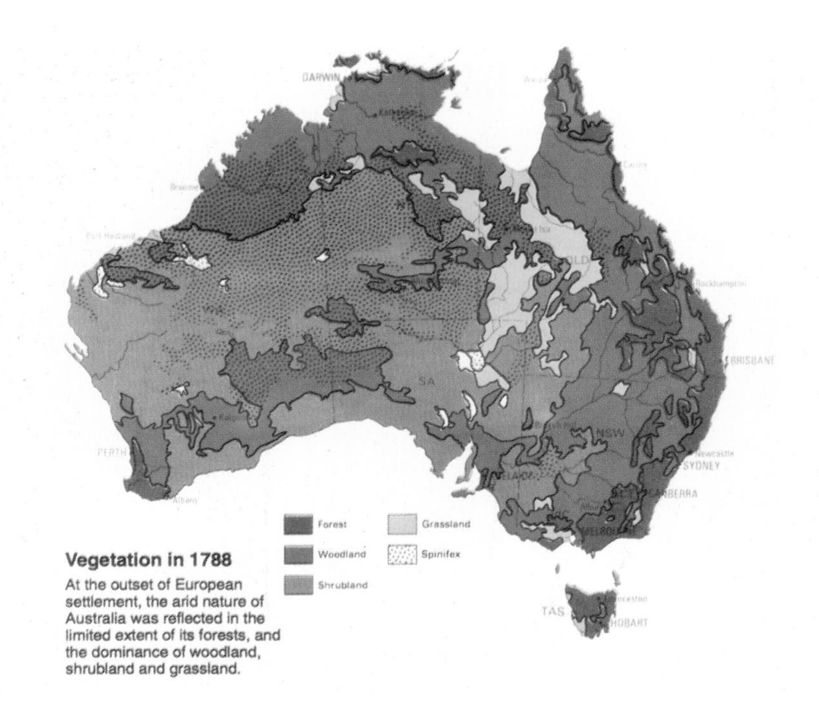

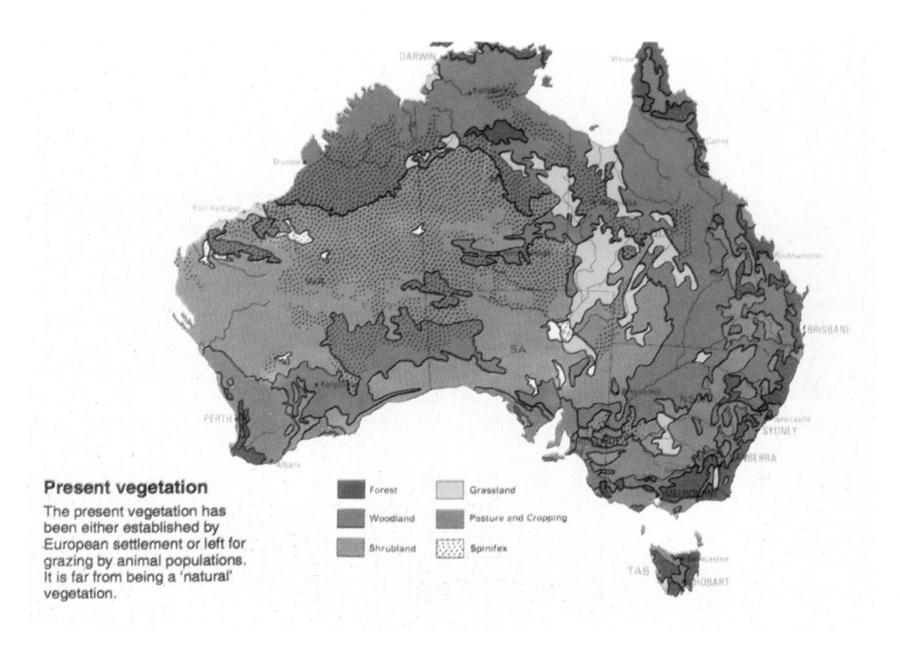

Plate C
Above and Right: Vegetation in Australia in 1788 and 1988, illustrating changes over this period. *(Source: ABS 1992.)*

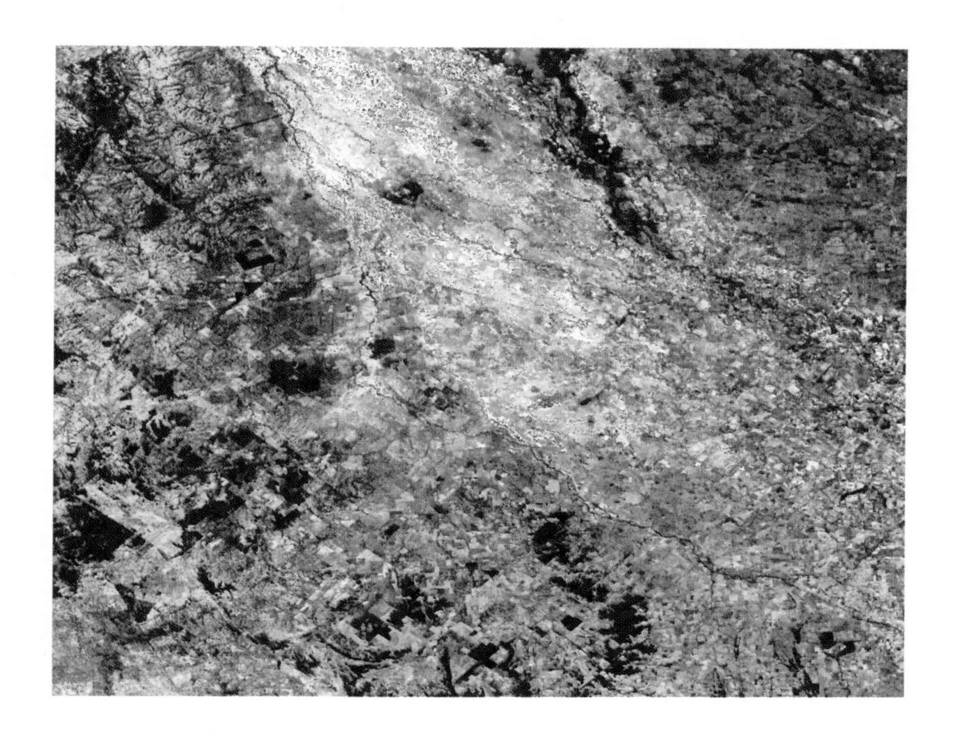

Plate D
Satellite images showing the clearing of poplar box country and the expansion of wheat fields in the Nyngan district, New South Wales, 1972–90.
(Source: Graetz et al. 1992.)

Plate E
Saline seepage leading to vegetation loss, and erosion extending close to a ridge line, south-western Western Australia.

Plate F
Straight drains cutting across the lower Shoalhaven floodplain in New South Wales and carrying acidic drainage into the estuary from acid sulphate soils.

Plate G
Woodchip stockpile, and conveyor from the mill, Eden, New South Wales.

Plate H
Forests near St Valentine's Peak, Tasmania, showing selectively logged native *Eucalyptus obliqua* forest in the background, *Pinus radiata* in the middle ground, and *Eucalyptus nitens* in the foreground. The pine and *Eucalyptus nitens* plantations are on former farm land. *(Source: North Ltd.)*

Plate I
The disused Yampire Gorge mine, the Pilbara region, Western Australia. The blue seam of asbestos can be seen in the cliffs. *(Source: photo by R. Delbridge.)*

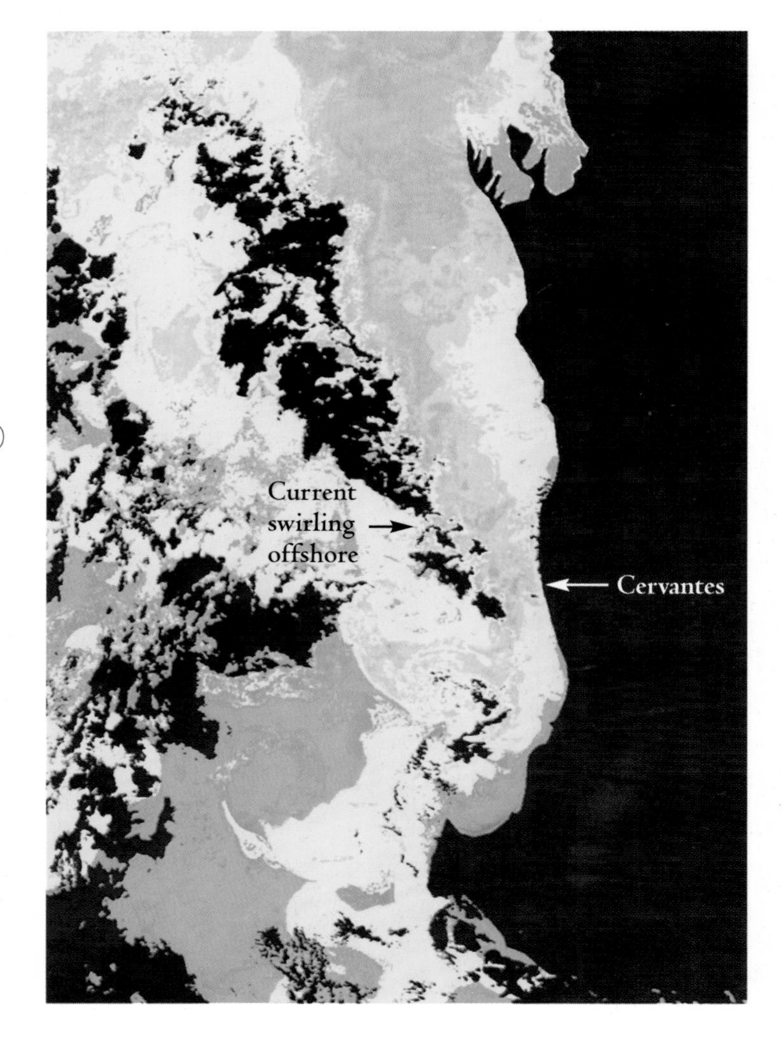

Plate J
Satellite (NOAA) image of the warm Leeuwin Current swirling offshore (red to yellow), at the time of the Kirki oil spill. *(Source: image supplied by WA Satellite Technology and Applications Consortium, processed by Remote Sensing Applications Centre, Department of Land Administration, Western Australia.)*

Plate K
Looking from beside the lock at Cullen Bay, Darwin, towards the marina. The high quality residential units demonstrate the value of waterfront housing sites in Australian cities.

Plate L
A 'red tide' of algal bloom off the coast of north Wollongong, New South Wales, in September 1994, probably as a result of a natural upsurge of nutrients.

Plate M
Mineral sand mine on Stradbroke Island, Queensland. The dark layer near the top of the dune is the indurated B horizon of the podzol soil developed in the dune sands. The mineral sand seam is just below the base of the pit. *(Source: photo by E. Bryant.)*

Plate N
Water polluted by garbage flowing across the beach at Port Kembla, New South Wales, after wave erosion breached dunes during storms in 1974.

Plate O
A house perched above the eroded bank of a confined channel in Wollongong, New South Wales, after the August 1998 flood.
(Source: photo by L. Mulvey.)

Plate P
Right bottom
The extensive irrigated cotton fields near Narrabri, New South Wales. Note the dark cracking clays of the fields, and the large on-farm dams.

Plate Q
Right bottom
The towers and ridges of the Bungle Bungle Range, Western Australia.

Figure 5.8
Cracks occurring across a sandstone 'pagoda' as a result of mining subsidence near Newnes, New South Wales.

Around the brown coal mines of Victoria's Latrobe Valley, the ground has subsided up to 1 m as a result of water extraction. Mines that supply the power stations to generate 75% of the State's electricity are open-cut pits, and the groundwater that seeps into them is extracted so that the sidewalls remain stable. This dewatering has lowered the level of groundwater within more than a 40-km radius of the mine because the rate of dewatering is about four times the annual recharge rate of the groundwater resource. Surface water has also been affected as the Morwell River has been diverted to allow mining of good quality coal below it (Department of Resources and Energy 1983).

At the Olympic Dam project in South Australia the use of groundwater has caused concern about the environmental effects of drawdown. This project involves the mining of a complex ore body (of uranium, gold, copper and silver) with on-site processing to produce metals and uranium oxide, and has an associated township. The area is arid, and the local groundwater is unsuitable, so water is taken from wellfields in the Great Artesian Basin, 110 km away. Drawdown is required to be kept at less than 2 m over a 30-year period, so that natural springs continue to flow (Waterhouse and Armstrong 1991). These springs deposit mounds of clays and various salts, and provide the habitat for a range of fauna (fish, aquatic snails and ostracods) and plants. However, many springs have been severely degraded throughout the Great Artesian Basin by cattle trampling them when they have been used for stock watering.

Australian mining and international issues

Mining in Australia has often relied on overseas capital and expertise for its success. Since the 1960s it has been affected by worldwide trends in environmental management, and Australian mining companies operating overseas have been influenced by environmental standards in Australia. As J. Wright has pointed out, the battle over mining in the Great Barrier Reef can teach us much about the relationship between Australian mining and international issues:

> The Great Barrier Reef has killed many ships and men over the years; but it has drawn and fascinated everyone who has ever seen it, and men have fallen in love with it as well as despoiled it. The story [of the battle against oil drilling and limestone mining on the Reef] will say little of the Reef itself. It is a political story, but also it is a story of people in their interactions. Its complications stretch far beyond Australia . . . and [into] the whole complicated structure of the world, its industry, its science and its commerce.
>
> Wright 1977, p. xiv

In the late 1960s plans to mine limestone and to begin oil exploration on the Great Barrier Reef were opposed vigorously by conservation groups. They sought the support of international groups, particularly the World Wildlife Fund; they involved Commonwealth politicians and the union movement; they tried to mobilise public opinion against the overseas companies proposing oil drilling. The limestone mining application was refused in 1969, and five years later a Royal Commission recommended a moratorium on oil drilling. The Reef was brought under the joint control of the Commonwealth and Queensland governments via the establishment of the Great Barrier Reef Marine Park Authority in 1975. But it was one of Australia's international obligations that finally ensured protection of the Reef. In 1981 the Reef was inscribed on the World Heritage list, and today mining is not permitted under the management procedures.

The Reef is only one area where mining and declaration of World Heritage legislation have been in conflict. In 1983 the Commonwealth government passed the *World Heritage Properties Conservation Act*, using its power over international affairs. This Act allows the government to prevent any activities that will degrade the qualities that caused the areas to be listed on the World Heritage register; this is the case for both areas that are listed and, since 1988, areas that are being nominated. In the view of one outspoken mining executive, the World Heritage register 'has become in Australia, what it is nowhere else in the world, a device whereby a central government, within a federal constitutional system, can supersede the states in controlling land-use management' (Morgan 1993). New proposals for World Heritage nominations have caused concern also in the pastoral industry. For example, graziers on the Nullarbor Plain fear invasions of weeds and feral animals if pastoralism is discontinued, yet the spectacular caves below the Plain are a strong argument for protection of the natural landscape (Figure 5.9). Declaration of several areas being considered for nomination as World Heritage may conflict with mining interests: the Lake Eyre Basin, because of the importance of the Roxby Downs (Olympic Dam) mine, and the Ruddall River in Western Australia, where a major uranium deposit occurs in the national park and on Aboriginal land. The most controversial example has centred on Kakadu National Park and the Conservation Zone within it.

In 1978 the Commonwealth government announced its interest in creating a large national park at Kakadu, in the catchment of the South Alligator River east of Darwin. The Ranger Uranium Environmental Inquiry—called the Fox Report, after the Inquiry's chairman—had recommended that uranium mining

Figure 5.9 Abrakurrie Cave, under the Nullarbor Plain. (Source: photo by B. Moule.)

should proceed in the region, under strict supervision and control. An initial area (Stage 1) was listed as World Heritage property in 1981, and a second adjacent area (Stage 2) in 1987. An area adjacent to the southern boundary of the park (Stage 3) was more difficult because of the mineral resources within it. In 1986 a 'Conservation Zone' was excluded from Stage 3 of the national park to allow possible development of 'mining projects of major economic significance' (Resource Assessment Commission 1991a), meaning the gold/platinum/palladium resource at Coronation Hill. A major environmental inquiry (Resource Assessment Commission 1991a, 1991b) found that:

- the erosion and weed infestations in the catchment of the South Alligator River, due to feral buffalo wallowing in it, far outweighed any likely effects of the proposed mine
- the highest predicted sediment yield into the river from mining would be less than 0.3% of present mean annual sediment load
- the mine was not likely to significantly affect archaeological or biological resources outside its boundaries
- potential water pollution (by cyanide-contaminated waste water) could be controlled by a 'no release' system
- any development, even of a single and well-controlled mine, would nevertheless mean that the ecological integrity of the national park was not maintained
- the Jawoyn Aboriginal people believed in *Bula*, an ancestral being who was associated with the area and who should not be disturbed by mining.

This last point was the deciding factor (Galligan and Lynch 1992). Strong support from the Prime Minister for the Aboriginal position prevented the mine from proceeding, and the entire area of Stage 3 was listed as a World Heritage area. Since 1996 government policy has changed and approval has been given to mine the Jabiluka deposit. This has triggered protests related to land rights and to World Heritage protection.

But it is not only Australia's obligations under the *World Heritage Properties Conservation Act* that is at issue. Major Australian mining companies, such as BHP, and multinational companies with substantial activities in Australia, such as CRA, are involved in the development of mining overseas. Their participation in the gold and copper mines at Bougainville and at Ok Tedi in Papua New Guinea means that the environmental impacts of those mines are assessed by Australians. The Ok Tedi mine is in the catchment of the Fly River, which discharges into the Gulf of Papua and thus affects Torres Strait. Annual rainfall is a massive 9 m, the region is seismically active, and huge landslips in 1983–84 prevented the completion of tailings dams. Since then the tailings have been discharged into the Fly River, the companies arguing that they would have minimal impact because of the vast volume and natural sediment load of the Fly system (Lawrence and Dight 1991). Ironically, the relaxation of environmental standards at Ok Tedi came about partly in response to economic disruption when the people of Bougainville rebelled and forced closure of that mine. In 1987 the two mines had supplied 65% of Papua New Guinea's export income and 15% of gross domestic product, so the loss of revenue from Bougainville after 1989 was a severe blow (Thompson 1990). However, the Australian government is concerned about the danger of heavy-metal contamination of the waters and sediments of Torres Strait, and the potential impact on fisheries, on green turtles and dugong (which are traditional foods for Torres Strait Islanders and are also endangered species), and on the northern extremities of the Great Barrier Reef (Lawrence and Dight 1991).

Relationships between Australia and the nation of Nauru, a neighbour in the central Pacific, have been strained by disagreement concerning responsibility for rehabilitation of mined areas. The island of Nauru is small—only 22 km^2—and has a very low average rainfall (2064 mm/year). Since 1906 phosphate has been mined on the island and exported, mainly for use in Australia as fertiliser. Mining leaves behind a wasteland of coral pinnacles 4–8 m high, at a density of 3–4 pinnacles per 100 m^2. After the First World War the island was administered under a tripartite mandate by Australia, Great Britain and New Zealand. When Nauru became independent in 1968, it bought full control of its phosphate industry from the three other governments (Manner et al. 1985). Nauru has since sought compensation from Australia for the environmental damage caused by phosphate mining. Australia has not accepted this claim, pointing out that only one-third of the area was mined prior to independence, and the Nauruans have mined the remaining two-thirds since then (Government of Australia 1993). A settlement of A$107 million was agreed upon in August 1993.

In one of its territories, however, Australia has been an international leader in the prevention of mining. This is in Antarctica, the world's driest, windiest and coldest continent (Lovering and Prescott 1979). International law accepts that Antarctica is *terra nullius* and thus subject to national sovereignty, although sovereignty claims are not formally recognised. Australia has claimed sovereignty over parts of the continent, following their transfer from the British government in 1936 (Figure 5.10). This sovereignty is usually established by exploration and continuing scientific research. Following the International

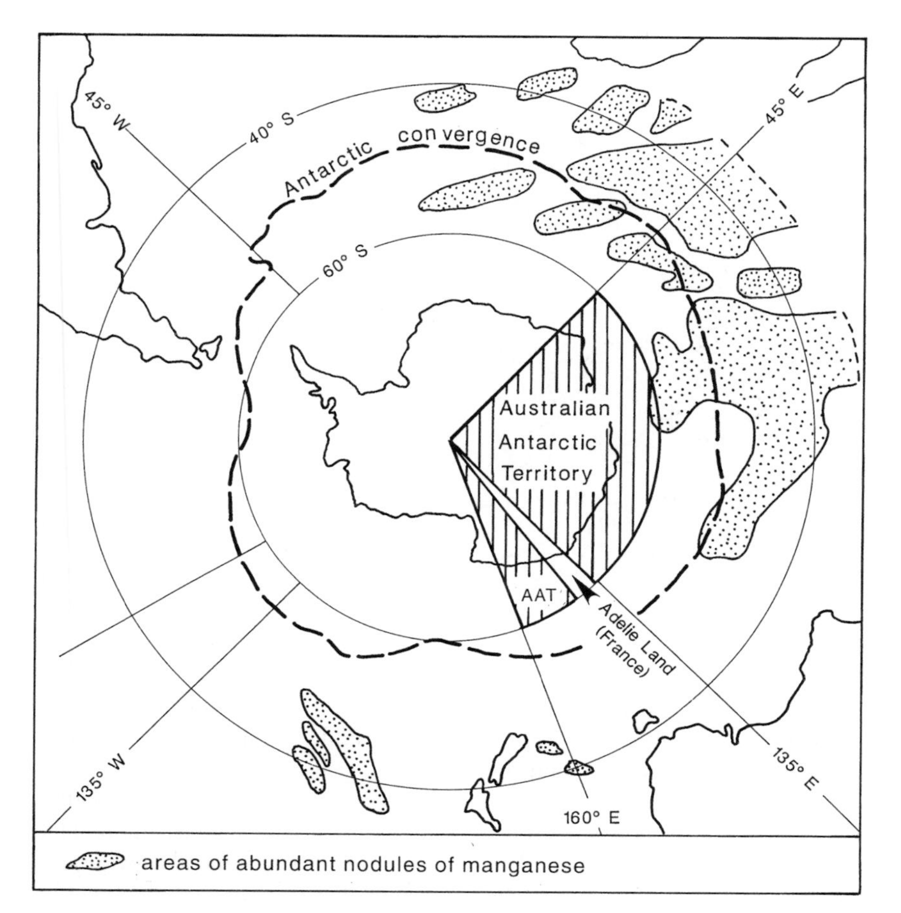

Figure 5.10 The Antarctic region, showing the Australian Antarctic Territory and offshore manganese deposits.

Geophysical Year of 1957–58 the 12 countries claiming Antarctic territory agreed upon a treaty to maintain cooperation, preserve the continent as a 'laboratory' and ensure its use for peaceful purposes only. In 1972 world conservation groups called for it to become a 'world park'; in 1977 a moratorium was called on all mineral exploitation and exploration (Davis 1992, Senate Standing Committee on Natural Resources 1985). Some mineral deposits had already been identified, and the extent of mineral resources in formerly connected parts of Gondwanaland indicated a wealth of potential deposits (Figure 5.11). The moratorium was called in response to concern that even exploration was too damaging to the fragile environment of the continent. More recent evaluation suggests the minerals may not be worth retrieving. Much of the continent is below ice which can be 2.5 km thick, making both exploration and recovery of any identified deposits difficult and uneconomic. Coal and iron ore have been discovered; oil and gas reserves offshore may be present; but many other minerals are unlikely to be found. The offshore manganese nodule deposits are much less rich than those of tropical areas, and even those are not economic to mine under present conditions (Antarctic Division 1999).

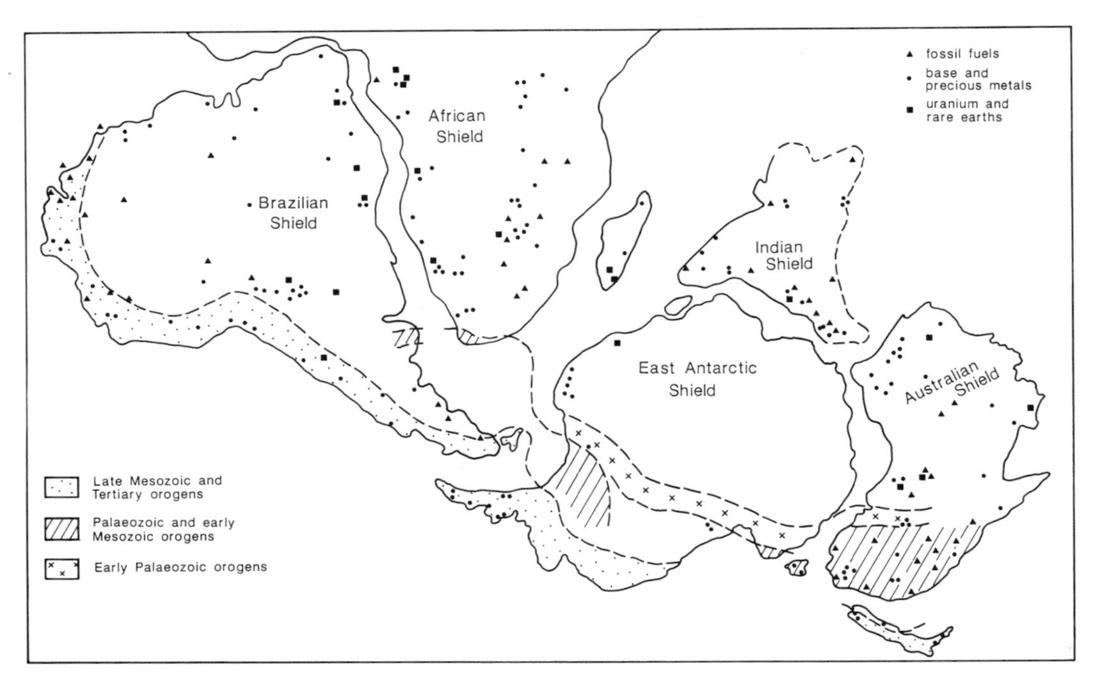

Figure 5.11
The distribution of known mineral deposits and major geological structures across Gondwanaland. (Source: data from Derry 1980, and Lovering and Prescott 1979.)

The Convention for the Regulation of Antarctic Mineral Resource Activities (CRAMRA) was proposed, which would have allowed some mining exploration and then development under stringent safeguards; but in 1989 Australia and France decided not to sign CRAMRA. In 1991 the Madrid Protocol on Environmental Protection in Antarctica was agreed upon. This came into effect in December 1998. This agreement bans mineral activities for 50 years and has such severe review conditions that the ban can effectively be regarded as permanent (Davis 1992). It also regulates environmental impact assessment, flora and fauna conservation, waste disposal and management, and prevention of marine pollution. The Antarctic islands—Heard Island, McDonald Islands, Macquarie Island—were added to the World Heritage list in December 1997.

J. Handmer and M. Wilder (1994) identify several significant factors in the Australian government's success in preventing mining in Antarctica:

- recognition that Antarctica's environment could only be protected through a comprehensive agreement, and not by piecemeal measures
- strong international support from non-government environmental groups such as Greenpeace, and the Antarctic and Southern Ocean Coalition (which represents over 200 organisations from 35 countries)
- political support within Australia for the government's position
- the use of the Precautionary Principle—one of the key principles of ecologically sustainable development (ESD) and the Intergovernmental Agreement on the Environment (IGAE)—since the Principle states that scientific uncertainty is not an excuse to avoid or delay environmental protection.

Conclusion

In the past the environmental impacts of mining have been severe. The example of Queenstown is particularly dramatic, but in many other places less obvious damage persists. Eroded areas, water-filled pits, weed-covered mullock heaps, diverted drainage, abandoned settlements, and sediment-laden streams all bear witness to an attitude within society that allowed extraction of minerals with no thought for the future. Today the industry takes a far more responsible approach and, indeed, argues that it is often at the forefront in developing rehabilitation and land-management techniques. However, the implications of mining extend well beyond the immediate boundaries of the mine itself. Transport and processing of the extracted materials, and provision of services for the associated workforces, also have environmental effects. For instance, Australia's minerals industry is strongly export-oriented and, as a result, relies heavily on sea transport. In the next chapter we will look at coastal and marine environments, and the implications of the major ship movements that service Australia's export industries.

Chapter 6

Marine and Coastal Environments

Introduction

THE SEAS are the 'Cinderella' of environments: neglected, poorly understood, exploited, but with a value and importance that is becoming recognised more and more. They are unlike terrestrial environments in that they are three-dimensional on a vast scale. Nutrients, food and energy are constantly being transferred across sites, both vertically through the water column, and horizontally over long distances in currents. However, in a land-based ecosystem, while energy and water may be introduced from the atmosphere, the system is largely self-contained in two-dimensional space, with only a narrow vertical band from soil to treetop. The life cycles of marine animals are difficult to observe and often unlike those of land animals. We admire most those which are most like us—whales, turtles, dolphins and seals, which care for their young, breathe air and sometimes have complex communication systems. We disregard or may be repelled by many other creatures, and imply that they are unattractive when we insult people by calling them 'slippery as an eel' or a 'cold fish'. Predicting the response of the seas to environmental change and resource use is as difficult as predicting the weather, with the added complication that its physical environment is even more difficult to observe. Hence, it is not practical to simply transfer management concepts that work reasonably well on land to the marine environment (Kenchington 1990). Marine ecosystems are more open and interconnected than terrestrial systems, so it is harder to protect marine reserves from outside influences. For all these reasons, the marine environment is poorly understood and inadequately studied, and its resources are often poorly conserved.

Poorly conserved resources

Australians have been far less concerned with conservation in marine environments than in land environments. Australia's first marine park (Green Island, off Queensland) was declared in 1938, over 50 years after its first terrestrial parks were proclaimed. By 1997, 38.9 million ha were included in marine protected areas, but the majority of this—some 34.8 million ha—lies in the Great Barrier Reef Marine Park. A further 430 000 ha are in Ningaloo Marine Park in Western Australia, and most of our marine bioregions have minimal or no protected areas. Nevertheless, Australia has 24% of the total number of marine parks worldwide (Zann 1995). The marine protected areas occupy about 3.5% of the Exclusive Economic Zone, or EEZ (declared in November 1994 and formerly called Australian waters). This can be compared with the 58.6 million ha of terrestrial protected areas, which occupy about 7.6% of the Australian continent (Cresswell and Thomas 1997). The area under Australian jurisdiction offshore is comparable to the area on land, and, like the terrestrial environments, contains unique and diverse biota.

Ironically the part of the marine environment that is most familiar to us—the coastal margin—is, in fact, atypical of the seas, in that land-based processes impinge on it so strongly. Yet we become aware of environmental impacts on the marine environment mainly because we observe their influence at the coast, as oil-affected birds wash up onto the beaches, as fish become scarce in city markets, as oyster farms are affected by toxins from algal blooms after release of contaminated ballast water from ships, or as tourism to offshore reefs promotes urban growth on the coast.

Unfamiliar and unappreciated though it may be, the sea is vital to Australia's historical development and present economy. Most Australian exports and imports move by sea, and so transportation is a major use of the marine and coastal environment. Building of civil and naval ships, and boats, boosts both domestic productivity and export income (McKinnon 1993). Offshore and nearshore mineral reserves—particularly of gas and oil—are both economically and environmentally significant (Figure 6.1). As on land, commercial exploitation of resources competes strongly with conservation of the natural environment, and with recreation and tourism in some regions—notably the Great Barrier Reef. Fishing is unusual in that commercial and recreational users share the resource, and it is the only significant hunting industry in Australia.

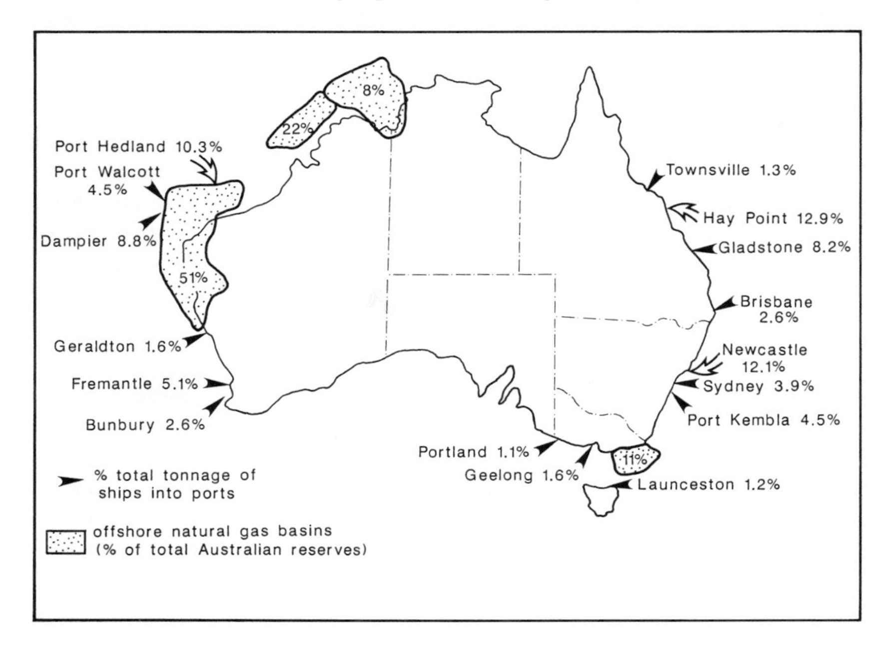

Figure 6.1
Major ship-entry ports and the location of offshore gas reserves around the Australian coast. (Source: data from ABS 1992.)

Commercially, fishing is less valuable than the major agricultural industries, but more valuable than sheep, pig, and poultry meat production; it yields about A$1137 million per year. Some A$3000 million is spent on recreational fishing annually (Ecologically Sustainable Development Working Groups 1991a) but its catch tonnage may only be about 25% of commercial finfish levels (Dovers 1994). Also, more recreational than commercial activity is probably focused on inland freshwater fishing, rather than on coastal and marine fishing. No data are available on fishing by indigenous peoples, an activity mainly taking place in northern waters. However, the importance of the marine environment to many indigenous people is great, and the concerns of the Torres Strait Islanders have led to a specific regional strategy for the Strait (Mulrennan 1993).

In world terms, Australia's marine environment is not highly productive. There is little runoff to the ocean, and Australia's rivers carry low nutrient loads because of the aridity and long weathering history of the continent. Unlike other continents in the Southern Hemisphere, there is no dominant upwelling current bringing nutrient-rich water to the western coast (Box 10). Although the EEZ is the third largest in the world, its commercial catch ranks 55th (Kailola et al. 1993). Even so, best available estimates suggest that there is little potential for expansion and that present catch levels are not sustainable. Of the marine fish, 25 species are overfished, 101 are fully fished and only 31 are underfished; the status is unknown for 70 species (Ecologically Sustainable Development Working Groups 1991a). Aquaculture of high-value products is becoming increasingly important, with oysters, pearls, Atlantic salmon, trout and ornamental fish being the dominant products. These industries have environmental implications: competition for estuary sites with tourism or conservation, nutrient-rich or chemical contaminated wastes, and potential introduction of exotic pests or diseases. However, unlike land-based agriculture, many of the species cultivated are not exotic, but are native to Australian waters (Kailola et al. 1993). Also unlike land-based industries, the fishing industry operates largely in areas under Commonwealth, rather than State or Territory, jurisdiction (Table 6.1) because Commonwealth jurisdiction extends from 3 to 200 nautical miles offshore (Figure 6.2). Furthermore, the resources themselves and the area exploited are publicly and not privately owned; thus governments have special responsibility for the management of the fishing industry and other marine resources (McKinnon 1993). However, they may still affect native species by loss of genetic diversity, culling of predators such as sea birds or seals, or demand for wild fish such as pilchards for food stock (Zann 1995).

Table 6.1
Commercial fish production (by percentage of gross value in dollars), 1995–96

	NSW	*Vic.*	*Qld*	*WA*	*SA*	*Tas.*	*NT*	*Cwth*	*Aust.*
Fish	7.8	5.3	14.2	9.9	12.3	13.8	2.3	34.5	25.8
Crustaceans	4.0	2.1	20.2	38.5	12.3	6.6	0.6	15.6	46.0
Molluscs	8.8	9.1	5.1	42.2	6.9	14.3	9.2	4.3	28.2

Data calculated from ABS Year Book 1999. Percentages include an estimated value for 23% of total production from aquaculture

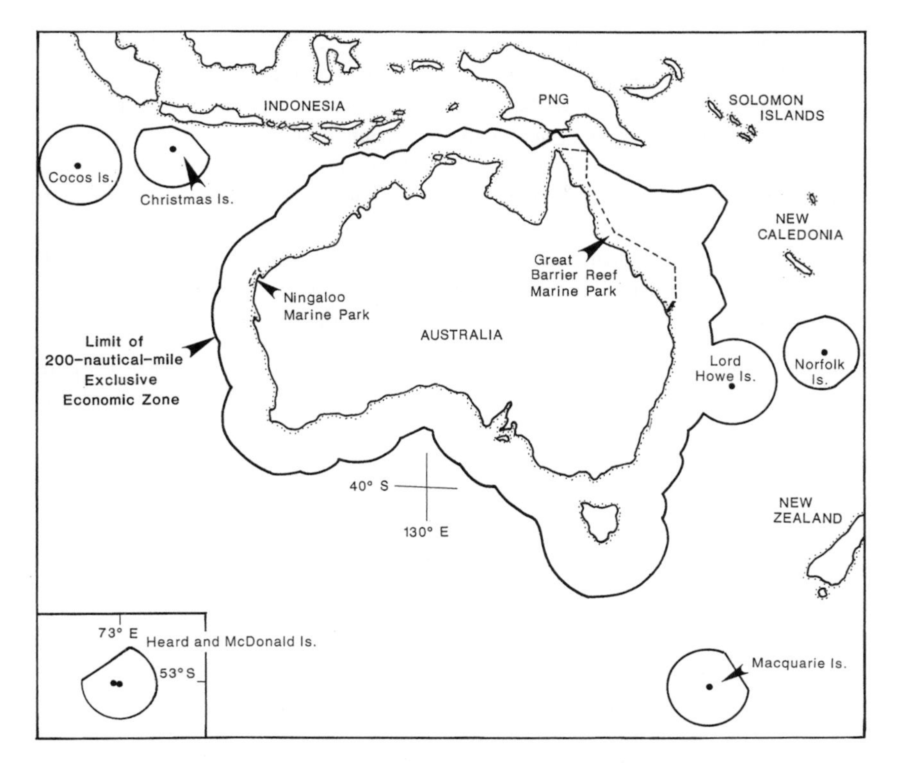

Figure 6.2 The 200-nautical-mile Exclusive Economic Zone and large marine parks. (Source: data from ABS 1992, and Zann 1995.)

Impacts on, and due to, fishing activities

Much of the current debate about management of fish resources focuses on problems of measuring and predicting the resource and its sustainability, and on allocating property rights, but there is little available information about environmental impacts on fish stocks, except in coastal areas (Brodie et al. 1990, Ecologically Sustainable Development Working Groups 1991a, McLoughlin 1992). A range of impacts can be identified along the coastal margin:

- destruction of habitat, particularly for breeding fish, by land reclamation or pollution of mangroves and seagrass beds in estuaries. Sewage and sediment from urban runoff has severely depleted seagrass beds in Cockburn Sound (Western Australia), Westernport (Victoria) and the Gulf of St Vincent (South Australia). In Cockburn Sound and Westernport, 97% and 71% of the original seagrass beds have been lost respectively. Dredging for harbour and airport construction, oil pollution and land reclamation has also destroyed some of the mangroves and seagrass beds in Botany Bay (New South Wales)
- release of toxic substances into coastal waters. Drainage from acid sulphate soils has caused fish kills in rivers in northern New South Wales and has damaged prawn stocks in the Shoalhaven River (see Box 5). Some substances, such as pesticides, are toxic. Others, such as putrescible organic wastes, can reduce dissolved oxygen to lethal levels
- changes in the flow regime or sediment loads of coastal streams. These changes can affect fish habitat, again particularly in the estuaries used for

breeding. Barrages on tidal reaches of streams in central Queensland, constructed to ensure fresh water supplies to nearby towns, have been blamed for loss of barramundi stocks because they interrupt upstream migration and, thus, the breeding cycle. Dams upstream may reduce sediment loads and associated nutrient loads to the estuaries and nearshore areas. Conversely, high sediment loads from urban and agricultural areas may smother some environments or cause problems of eutrophication

- pollution of mariculture areas. Oysters from leases in estuaries near major cities such as Sydney have been contaminated by pathogens from sewage overflows during wet weather and have been made unsaleable.

Fishing may itself cause adverse impacts, both in coastal areas and in open ocean (Brodie et al. 1990, Ecologically Sustainable Development Working Groups 1991a, Wace 1994, Zann 1995):

- Trawl nets and shellfish dredges can scour into the bottom sediments of bays and coastal waters. This can create scour channels, cause smothering of sedentary life forms as the sediments resettle, and physically remove some benthic life forms. The effect varies in different environments. In Jervis Bay, New South Wales, where the sea bed is sandy, trawling seems to have had minimal effect; but on the North West Shelf, off Western Australia, removal of sponges seems to have disrupted the ecosystem and led to reduced fish catches. Zoning prohibits trawling in parts of the Great Barrier Reef.
- The use of boats pollutes waters with waste fuels and highly toxic biocides, which are used as anti-fouling agents.
- Species other than the fish hunted are often caught inadvertently and subsequently die. There has been international concern about the catching of dolphins in gill nets, and of albatross and other sub-breeds on long lines. The 1989 meeting of the South Pacific Forum nominated drift netting as a priority area for international action because of its potential to fish out stocks and its high levels of non-target catch. Trawling for prawns captures large numbers of small fish and, in tropical waters such as the Gulf of Carpentaria, sea snakes and turtles. New net designs are being introduced to help unwanted fish to escape, but Australia has lagged in this regard.
- Another problem is the entanglement of marine animals, especially seals, in discarded fishing gear. A survey of litter along a secluded beach on the Western Eyre Peninsula in South Australia found that the debris came overwhelmingly from fishing activities on the nearby continental shelf. The debris included almost equal weights of hard moulded plastics (for example, drums, buckets, crates, craypots and floats) and flexible plastic (ropes, nets, lines and strapping).
- Recreational fishing can affect fish stocks. Gathering of seaweed, shellfish, sea urchins, cunjevoi and other fauna from rock platforms near urban areas can deplete these areas rapidly. A study in New South Wales showed an average of 55 people visited rocky shores per km per day, and about 15 of those 55 harvested biota. Fishermen taking cunjevoi (*Pyura stolonifera*) cause a chronic and extensive disturbance. Often animals such as sea urchins, abalone and shellfish are taken because they are regarded as delicacies in

many cultures. Removal of grazers like limpets and urchins can allow algae, weed and barnacles to spread, changing and simplifying the shoreline ecology. In gathering from the rock platforms, often the largest individuals of biota are taken, yet they are the most important for continuing reproduction of the species. The few marine-protected areas are poorly policed. Elsewhere control is sought by bag limits, but these are ineffective and, even if observed, would not prevent damage such as that caused by the use of crowbars to break under ledges (Underwood 1991). In 1995 the estimated recreational catch was 50 000 tonnes. In comparison the commercial catch was 195 000 tonnes, and the aquaculture production about 20 000 tonnes.

Damage by discharge: Oil and ballast water

Of all forms of marine pollution, oil spills are the most dramatic and newsworthy. The huge *Exxon Valdez* spill, which spread over 200 000 tonnes of oil along the Alaskan coast in March 1989, and its effects on the fishing industry there, continue to make headline news throughout the world. A spill during unloading from the *Laura D'Amato* to the Shell refinery in Sydney Harbour in early August 1999 led to calls to ban oil tankers from the harbour. In Australia incidents are monitored and dealt with by the Australian Maritime Safety Authority's national plan, which is funded by a levy on commercial shipping. In 1989 the Authority recorded 177 oil spills around Australia, although only 23 of these led to legal action being taken. Most took place in Queensland (82), many in Victoria and New South Wales (40 and 30 respectively), and fewer in the other States. If, however, the volume of oil spilled and the volume of dispersant used is taken as a measure of the problem, New South Wales, Victoria and Western Australia were the worst-affected areas, then Queensland, then the other States (note that volume data are for 1984–85) (ABS 1992). Offshore exploration and production activities are another source of risk but to date there have been no major incidents.

The Great Barrier Reef is one region where oil spills could cause severe damage, and oil/gas exploration is banned there and also in Ningaloo Marine Park (Western Australia). Every year 200 tankers carrying over 100 000 tonnes of oil, and 1800 ships with up to 5000 tonnes of fuel oil, travel along the Inner Route between the coast and the outer Reef. A large spill could not move out to open ocean, and limited wave activity means that it would not disperse well by natural agitation. It could ruin the reef life and the beauty of the reef over large areas. There were eight groundings and four collisions of large ships on the Inner Route between 1984 and 1990 (Preen 1991). It can be argued that the limited wave activity means that a large spill is unlikely because a ship is unlikely to break up, but since October 1991 pilotage of large ships has been made compulsory in an attempt to prevent collisions and groundings (Gray 1991).

While an oil spill remains in open ocean, its ecological effects are limited. The effect of the spill from the *Kirki* off Western Australia in July 1991 was minimal because the currents moved the oil away from the shore (see Box 10). Oil on the sea surface can be evaporated, chemically broken down and biodegraded, and these processes are accelerated when it is dispersed by wave action. Oil does not sink readily and so does not easily enter the food chain of marine fauna (Figure 6.3). Near the coast, however, oil is a greater problem. The effects of an oil spill on a coastline are unsightly, and public sympathy is rightly aroused by the sight of animals with their bodies clogged by oil. Seals and birds are very susceptible to damage because the oil affects their body insulation. Whales and dolphins are smooth-skinned and less susceptible, but can become ill by eating oily fish and molluscs. There may be damage to sensitive environments such as mangroves and seagrass beds, which are the breeding or feeding grounds of migratory birds protected under international treaties. Control of spills in these shallow areas may be difficult if surf or currents preclude the use of containing booms, or if the spill is a substance, such as gasoline or kerosene, that cannot be contained because of the risk of fire. Dispersants should not be used in sensitive environments because they may be directly toxic and also because they spread the oil down through the water to the bed (State Pollution Control Commission 1991).

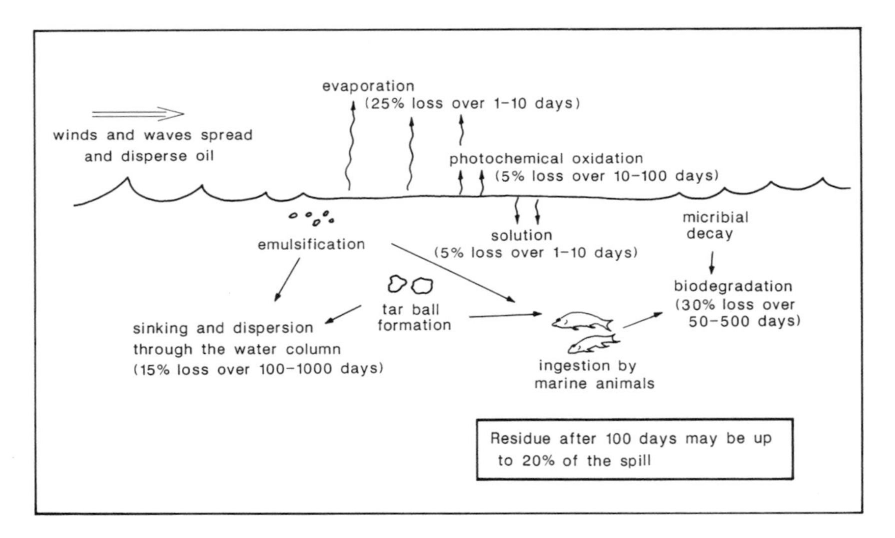

Figure 6.3
The fate of a crude oil spill. (Source: data from Bishop 1983.)

Oil can remain an obvious pollutant long after it is spilt. When Thor Heyerdahl sailed over 3000 nautical miles from Morocco to Barbados, he observed tar balls and oil slicks in the middle of the Atlantic Ocean. Oil is released by accidents, but also by tanker ballast discharge, from natural seeps and from oil-well leaks. It drops out from the atmosphere when vehicle and industrial emissions are blown out to sea. And, importantly, it washes into the sea from rivers and creeks. Some researchers believe that the volume of oil entering the sea from runoff is comparable to the volume resulting from marine transportation (Bishop 1983). By regional standards, Australia's coasts are not heavily polluted by oil (Figure 6.4), and much of the oil along the coasts comes from urban runoff or sewage discharges, but the consequences of an individual oil spill can be so severe that the matter is nevertheless of great concern.

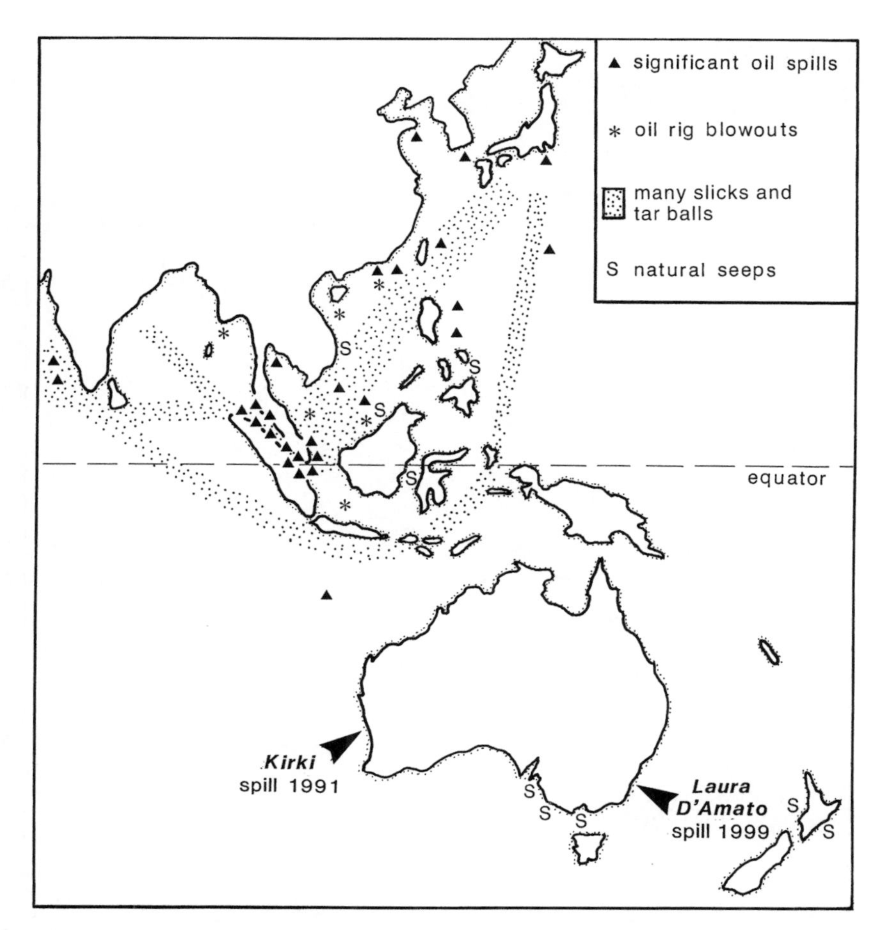

Figure 6.4 Marine oil pollution in the Australian–South East Asian region, showing significant spills in SE Asia 1974–89 and in Australia since 1990. (Source: redrawn from Couper 1990.)

Box 10
The Leeuwin Current

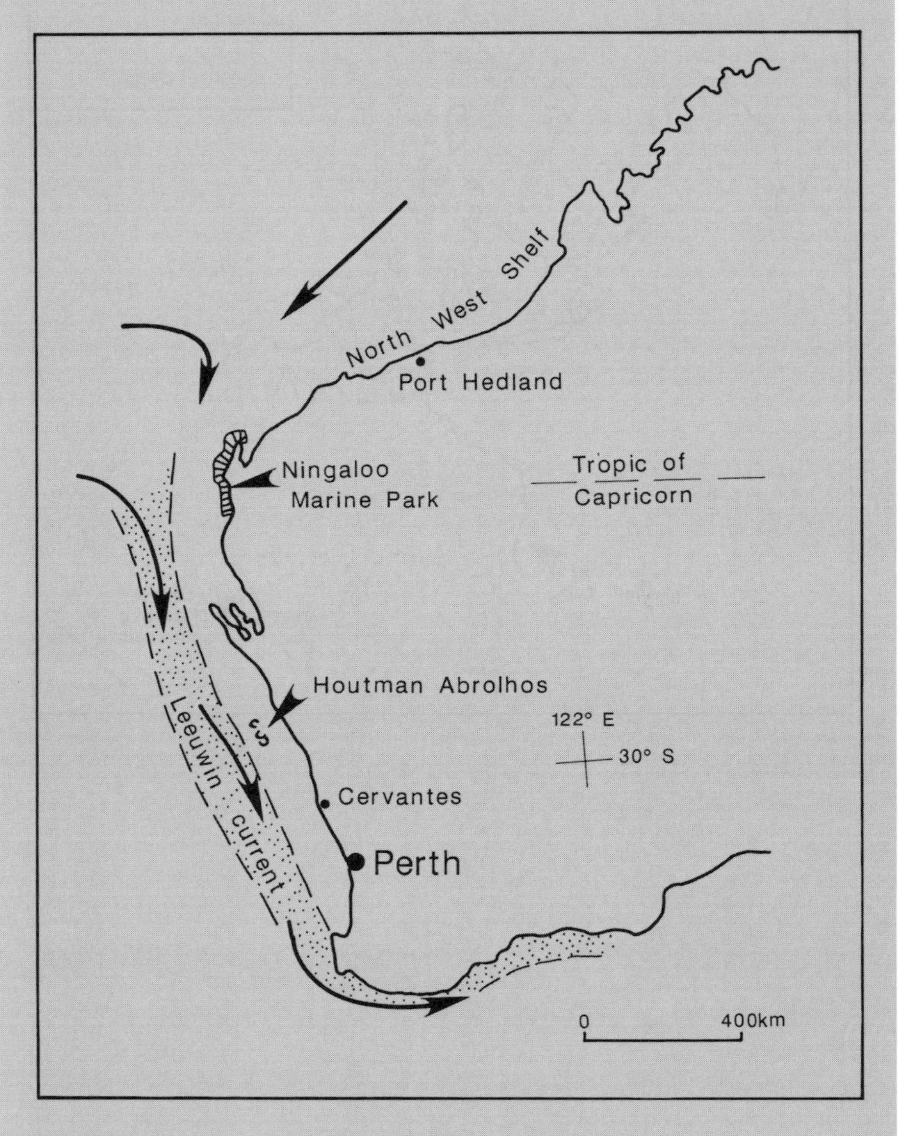

Figure 6.5
The Leeuwin Current moving southwards past Ningaloo Marine Park. (Source: data from Australian National Parks and Wildlife Service 1990.)

The west coasts of southern Africa and South America have strong north-flowing cold currents (the Benguela and Humboldt currents), which are upwellings of nutrient-rich water and which support productive fisheries. Western Australia is different. A south-flowing warm current—the Leeuwin Current—dominates its coastline from near Exmouth down to Cape Leeuwin and from there eastwards into the Great Australian Bight (Figure 6.5). It is responsible for some upwellings, but is important mainly in bringing warm water southwards. The current is 50–100 km wide and 100–200 m deep; its water is less saline (by about 0.2 parts per thousand) and warmer (by 1 °C) than the ocean water outside it. It is strongest in autumn and winter, and it brings tropical water down to the temperate coastline. While the current runs along the edge of the continental shelf, eddies swirl towards the coast. It is this warm flow from the tropics that supports the coral reefs and reef fauna as far south as Rottnest Island, near Perth. In July 1991 an oil tanker, the *Kirki,* lost its bow and released 16 000 tonnes of crude oil into the sea about 20 km off Cervantes, midway between Perth and Geraldton. The ship was damaged in the rough seas and 25-knot winds associated with an intense low pressure system. A major environmental problem was averted partly by salvage of a further 67 000 tonnes of oil from the tanker and because a tug pulled the tanker further offshore, but primarily by a huge eddy in the Leeuwin Current. This eddy swirled the oil away from the shore, where rough seas broke up the slick (Plate J).

Ningaloo Marine Park lies between North West Cape and Amherst Point, in the area where the current first becomes clearly defined. The park occupies 4300 km^2 and was declared under joint State and Commonwealth national parks legislation in 1987. It is a coral reef 260 km long and lying up to 7 km offshore. It is an important habitat for whales, whale sharks, billfish (such as marlin), dugong, turtles and dolphins; its waters play a significant role in tuna migrations; over 25 species of birds listed under international migratory bird treaties use the area. At present, commercial fishing is limited, and the park is used mainly for charter fishing and for tourism. Large numbers of whale sharks congregate when the corals spawn in April and May, and the chance to see these attracts many visitors. The washing ashore of litter, particularly plastics, is a significant problem, but a new international agreement on reducing pollution from ships may help to alleviate this.

The coral and limestone reefs washed by the Leeuwin Current between Shark Bay and Cape Leeuwin support Australia's most valuable single species fishery, the west-coast rock lobster fishery, from which A$200 million worth of catch is taken annually. Most is exported. Currents moving offshore in summer are important in carrying lobster larvae out to the ocean, and other currents return them, between nine and 11 months later, to the reefs, where they settle, moult and develop. The reefs like the Houtman Abrolhos differ from most coral reefs because they have extensive seaweed beds. Nutrients from the weed are released into the water when waves break off kelp and sargassum, and this nutrient release seems to be important in supporting the lobster fishery. There must be some source of nutrient renewal to replace the nutrients exported from the ecosystem by the removal of the lobsters. The fishery is closely managed. Research has allowed lobster populations to be predicted four years in advance, and no amateur fishing is allowed.

Data from Australian National Parks and Wildlife Service 1990, Ecologically Sustainable Development Working Groups 1991a, Gray 1991, Saueracker 1991, *Sydney Morning Herald* 22 July 1991, *West Australian* 5 August 1991

Release of ballast water from bulk carriers and other ships may be a less obvious form of pollution, but its impacts in the long term are likely to exceed the impacts of oil spills. The largest tonnage of ships comes into Australia at Hay Point and Gladstone in Queensland, Newcastle in New South Wales, and Port Hedland and Dampier in Western Australia (see Figure 6.1); over 50% of these ships come from Japan. The woodchip carriers entering ports in the southern States come without cargo and ballasted with water, and other carriers also carry large volumes of ballast water (on average about 60% of dead weight tonnage); on entry to ports for loading, this ballast is discharged into Australian waters. The ballast water usually contains some sediment, and unwanted organisms may be present in both the water and the sediment. A recent survey identified 16 species of fish, invertebrates and algae that have been introduced to Australian waters by ballast discharge (Jones 1991). Voluntary guidelines were introduced in 1990 to reduce ballast discharge and to reduce the sediment levels in ballast, but there is still concern.

The possibility of introducing exotic species into Australia's expanding mariculture areas is high, since many ships come from ports where extensive mariculture is practised. Mariculture organisms are very susceptible to infection and diseases because they are at much higher stocking densities than uncultivated animals and plants. Hence, imported exotic diseases could spread rapidly—and disastrously—through mariculture areas. Commercial shellfish operations (for oysters and mussels) in the Derwent and Huon estuaries in south-eastern Tasmania have been closed periodically since 1986 because of a species imported in ballast water. This is the dinoflagellate *Gymnodinium catenatum.* In its spore or cyst phase, this organism rests in fine sediment and can be taken up into ballast water, then transferred to the sediments of another port by ballast discharge. It then enters a motile phase, during which it moves near the water surface and can be ingested by filter-feeding molluscs like oysters and mussels. Unfortunately this phase is toxic to people and causes paralytic shellfish poisoning. When the water is calm, warm and high in nutrients, the motile phase can bloom, forming a 'red tide' of toxic organisms and forcing the closure of any shellfish farms nearby. In March 1999 the black striped mussel (*Congeria sallei*) was detected in a Darwin marina, and a massive operation to remove it involved chlorination and the use of heavy doses of copper sulphate. The outbreak could be controlled only because the marina at Cullen Bay is isolated by a lock and sea wall (Plate K).

Nutrient enrichment, runoff and sewage

Under natural conditions most of the ocean is nutrient-poor, and this is so even along most of the shallow areas of Australia's coastline. Thus, our coastal areas are very susceptible to eutrophication as a result of enhanced levels of nitrogen, phosphorus and other nutrients. As an algal bloom grows, the dissolved oxygen content of the surface water can be very high due to high rates of photosynthesis. When the bloom decays, however, oxygen is consumed and can fall below the level of saturation necessary for aquatic life. In tropical waters the effects of eutrophication are enhanced, because warm water has a lower capacity to hold dissolved oxygen than cooler water. While algal blooms are a natural phenomenon (Plate L), they may be more common because of increased nutrient loadings. In Cockburn Sound, enrichment by nutrients derived from urban and industrial wastewater led to the destruction of most of the seagrass beds in the late 1970s (Cosser 1997).

Australia's most extensive bloom in an embayment began almost overnight in December 1992 in Jervis Bay, New South Wales. It persisted for five weeks, and was followed two months later by a bloom of another alga extending from Forster to Port Kembla. Many people blamed sewage discharge into Jervis Bay for the first bloom, and subsequently the Shoalhaven Council adopted a preferred option for on-land disposal of effluent except under extreme conditions. Research has allowed other blooms to be predicted when several factors combine—offshore upwelling of nutrient-rich cold currents from the Pacific, a prolonged period of warm calm weather with flat seas, and some rain. It has not yet defined the role of nutrients from onshore sources (Eckersley 1998). It is not a simple matter to define this role, and results from one coastal area may not be applicable to others. Eyre and France (1997) comment that most studies are being done in temperate estuaries with disturbed catchments and/or sewage effluent discharges. Their study is based on a subtropical barrier estuary with a flat coastal plain behind it—the Pumicestone Passage behind Bribie Island in south-eastern Queensland. Phosphorus inputs to the Passage are closely linked to sediment inputs, and come predominantly from offshore. The marine source dominates because of low erosion in the catchment. In contrast, nitrogen comes mainly from the catchment, from leaching out of agricultural and horticultural areas, and apparently via groundwater in dry periods. Thus, in this case, regulating nitrogen yields from the catchment is more likely to improve water quality in the estuary than modifying phosphorus yields.

Recent research on the impacts of sewage discharge into marine environments has identified faecal sterols; organic compounds which are source-specific. Tracking sewage pollution by older methods, notably faecal coliform counts, did not allow sources to be discriminated. A study at Tuggerah Lakes, New South Wales, showed that up to 80% of faecal contamination in the lake water after rain came not from human sources but from sea birds occupying the lake shore. Another 15% came from domestic animals, most of the remainder from rural animals, and a negligible amount from human sewage (CSIRO Marine Research 1999). Similar techniques allow plotting of the sewage plumes from the deep ocean outfalls off Sydney, and geochemical signatures distinguish those plumes from estuarine outflows. It is accepted generally that sewage discharges affect marine communities, but documenting the impacts is difficult. Roberts (1996, p. 8) comments, after a study of macrobenthic communities near the North Head outfall, that 'considerable variation in the mean number of species and the abundance of most organisms was apparent on the deepwater reefs. This does not mean that there were no changes but rather that the natural fluctuations . . . may be greater than those caused by possible effects of sewage'.

The use of Antarctic waters

As noted in Chapter 5, the minerals of the Antarctic are now protected. The protection of the marine life has been a goal of the Antarctic Treaty and has been strengthened by two Conventions: one for the Conservation of Antarctic Marine Living Resources (signed in 1980) and another for the Conservation of Seals (covering seals in the ocean south of 60°S). These have been necessary because the history of exploitation in the region has been a sorry one. Hunting of seals in the sub-Antarctic from the late eighteenth century ended within 50 years because of the almost complete extinction of the fur seal and sea elephant. After whaling ships were fitted with cannon-harpoons from the early twentieth century, first the blue whales, and then the common and Rudolph rorquals, were hunted to drastically low levels. Attention has turned now to krill harvesting and the possibility of overexploitation of this resource.

The rapid exhaustion of resources occurs because of the peculiar nature of the ecosystem. Organisms are much larger, slower to mature and longer-living than their temperate and tropical counterparts. For example, krill—the invertebrates that feed on phytoplankton—reach a size of 2–6 cm whereas zooplankton in temperate seas rarely exceed 1 mm. Krill are abundant, with their biomass being perhaps six times greater than the biomass of zooplankton in the biologically rich North Sea. The krill may live for seven years, but temperate equivalents produce from ten to 20 generations per year. In addition, the physical environment is unusually rigorous. The whole food chain begins with algal microplankton, which are adapted to the severe cold but whose photosynthetic activity is limited by light availability. The high reflectance of the sun's radiation, due to ice and the choppy surface of wind-whipped water, allows very little light to penetrate the water near the ice fronts, where most of these algae are found. Hence, the ecosystem is very sensitive to disturbance, and overexploitation can cause permanent or, at best, long-lasting damage to the resources (Billen and Lancelot 1992).

An increase in the ozone hole in the stratosphere above the pole poses an indirect threat to the algal production. Release of inorganic chlorine compounds (CFCs) from refrigerants and aerosol spray cans are thought to reduce ozone levels, and therefore to allow more harmful ultraviolet radiation to penetrate the earth's atmosphere. If this occurs, algal production is likely to be affected (Stromberg et al. 1990).

The fragility of the marine ecosystem has been recognised in the decision of the International Whaling Commission to declare a whale sanctuary from 40°S, excluding the area within Chile's 200-nautical-mile fishing zone. This decision, made in March 1994, was strongly supported by Australia. Since, in the mid-1960s, the pesticide DDT was detected in Antarctic waters and biota, pollution of the Antarctic marine environment has been of international concern. Fortunately there is a strong physical barrier between Antarctica and its circumpolar ocean, and the oceans at lower latitudes. The Antarctic Convergence at about 50°S is a narrow zone where warm surface water flowing south mixes with the deep cold current that flows northwards and eastwards around the polar region. This circumpolar current isolates the Antarctic continent from pollutants carried by the warmer waters of lower latitudes. The pollutants that affect the marine environment are either produced locally (from research stations or fishing activities) or are transported in by atmospheric circulation. The pesticides detected in Antarctic waters like DDT or PCBs are derived from the atmosphere. Nevertheless, the levels of these are still in orders of magnitude lower than in the Northern Hemisphere, and Antarctica remains one of the least polluted parts of the globe (Stromberg et al. 1990).

Clean seas, dirty coasts?

On a world scale, the United Nations' Group of Experts on Scientific Aspects of Marine Pollution (GESAMP) concludes that the open ocean is still clean, but coastal regions are deteriorating. Not all researchers agree with this conclusion, and the data are so sparse and variable that disagreement is hardly surprising. W. J. Davis (1993) comments that PCBs (polychlorobiphenyls from pesticides) are more concentrated in open ocean than coastal waters, and that DDT is higher in ocean than coastal waters in the Northern hemisphere, but higher in coastal than ocean waters in the Southern hemisphere. Heavy metals are, however, highest near coasts, with 'hot spots' near major cities. Perhaps the whole discussion simply underlines the variability between different pollutants and between different places.

In the Australian situation it is difficult to argue with the claim that 'coastal ecosystems of the world bear the burden of most disposal of human wastes' (Underwood 1991, p. 167). Australia's large cities are coastal, and—except for the inland streams draining largely natural areas—its rivers carry industrial, sewage and agricultural wastes from their catchments to the sea. The identified 'hot spots' for contamination by metals, pesticides and hydrocarbons are from coastal areas near major urban and industrial complexes. While there is still cause for concern, evidence does point to an improvement in the levels of contamination since the 1970s, when environmental control legislation was introduced (Brodie et al. 1990).

In some cases, new channels are created to move pollutants more efficiently into the coastal zone. Eutrophication problems in the Peel–Harvey estuary,

70 km south of Perth, have been caused by high levels of nutrients from sewage and industrial waste, and from runoff from intensive agriculture in the catchment. There have been massive blooms of blue-green algae; decomposing algal slimes, offensive smells and fish losses have been major problems (Brodie et al. 1990). To allow greater flushing and export of phosphorus from the estuary, a channel is being cut between the estuary and the ocean.

In other cases, the estuaries bear the load of pollutants discharged well upstream. Discharges of heavy metals into the King River from the Mt Lyell copper mine at Queenstown have been blamed as the direct or indirect cause of fish kills in Macquarie Harbour in western Tasmania (Carpenter et al. 1991). Two contrasting streams flow into the Macquarie Harbour estuary in western Tasmania: the King River (17% of inflow), draining the mining region of Queenstown, and the Gordon River (83% of inflow), which is dammed upstream, but drains mainly wilderness area and national parks (Carpenter et al. 1991). In the Gordon River, and in the King River above Queenstown, the concentrations of most heavy metals are comparable to the concentrations in clean coastal waters. In Macquarie Harbour the story is very different because of the inflow of heavy metals from the lower reaches of the King River below the mining region. Most of the copper coming down the King River is in particulate form, but even the dissolved load is more than 50 times the concentration in the Gordon River. The particulate load falls out to the bottom near the mouth of the river, but changes in water quality (particularly low oxygen levels) can bring the metals back up into the harbour waters in solution. Thus the copper concentrations in harbour waters are from five to 16 times as great as in the Gordon River, and lesser concentrations can be traced outside the harbour as far as Cape Sorell. The situation is worsened because the levels of copper are higher than the levels of organic matter that could neutralise the copper's toxicity. Like most heavy metals, copper is toxic as a free dissolved ion. When it forms a complex with organic molecules (from decaying vegetation), it is not toxic to marine life. Unfortunately, the concentrations of copper in Macquarie Harbour far exceed the organic matter available to buffer its toxicity.

The effects of heavy metals on marine life may be a result of direct toxicity or of indirect effects on the life cycle. In Port Kembla harbour, New South Wales, discharges from the industrial complex around the steelworks included high levels of heavy metals, cyanide, phenolics and ammonia. These pollutants retarded the settling of the larval stages of bryozoans so that they remained as plankton for longer and were reduced in numbers by predation (Moran and Grant 1993). An anti-fouling agent—tributyl tin (TBT)—has been dubbed 'the most toxic substance ever deliberately introduced to the ocean' and its use is restricted now in New South Wales and Victoria. In Sydney it caused the shells of commercial oysters to become deformed. In Port Phillip Bay, Victoria, reaction to TBT has been monitored by the changes it causes to the sex organs of shellfish. Organic tin compounds like TBT cause imposex, the imposition of male sexual characteristics on female marine snails. The process is irreversible, and so the incidence of imposex provides a good indicator of long-term TBT pollution. In Port Phillip Bay it was found to varying degrees near all harbours and marinas, but was not found away from boating sites (Foale 1993).

The coastline: Meeting line of land and sea

For its inquiry into Australia's coastal zone, the Resource Assessment Commission (1993) used an arbitrary '50 km inland' to define the 'coastal zone', and then looked at census and other available data for statistical and catchment areas with 50% or more of their area in that strip. The results give an indication of the importance of the coastal zone to the Australian population and to environmental protection in Australia:

- In 1991, 86% of Australia's population lived in the coastal zone, mainly in metropolitan centres (61%). But the non-metropolitan coastal centres grew three times as fast as the overall Australian population between 1971 and 1991.
- Both domestic and international tourism is focused strongly on the coastal zone. As Bird (1988) comments, on a hot summer's day about two million people use our beaches—12% of the population using Australia's most popular recreational resource.
- Yet urban land occupies only about 2% of the coastal zone. One-quarter of the coastline is bare ground (mainly sand dunes), 8% is used for agriculture, and natural vegetation occupies the remainder. Of the total coastline, about 10% is held for conservation purposes and another 10% is vacant Crown land. Over half is in private tenure.
- Most of Australia's World Heritage areas include significant coastal areas: the Great Barrier Reef, Lord Howe Island Group, Shark Bay and Fraser Island, obviously; but also Kakadu, the Wet Tropics of Queensland, the West Tasmanian Wilderness National Park and the Australian East Coast Temperate and Subtropical Rainforests.

Thus, pressures on the coastal zone are localised, in areas of urban expansion and places of high value for tourism. This is obvious near the metropolitan centres, but is also true in less densely settled places. For example, the mangroves of tropical Australia are more extensive, diverse and biologically productive than the mangroves of the more populous temperate and subtropical regions of Australia. They have not been disturbed to the extent that the southern stands have, and are far less disturbed than their tropical equivalents in other long-settled and densely populated parts of the Indo-West Pacific region. Some stands in northern Australia are remote from settlement and are available for protection and conservation. In contrast, the stands near cities such as Darwin are under constant threat: they may be 'reclaimed' for building land, drained to eradicate insect pests, dredged to provide canal estates, subjected to excessive fishing pressure or degraded by polluted runoff (Figure 6.6). Unfortunately, the most diverse and rich stands are in the higher rainfall areas. Darwin and Cairns, each with an average rainfall of 1500–2200 mm per year, support stands with over 25 species of mangrove; near Broome, where the average annual rainfall is only 580 mm, there are only 13 species. Yet it is the wetter areas where there is greatest population and tourist pressure (Hanley 1992).

Canal estates near major urban centres are the result of increasing pressure for coastal building land, and for land with direct water frontage. They are often

Figure 6.6
Dredging and mangrove removal to provide for a canal estate and marina, near Darwin.

opposed by both fishery operators and conservationists because they greatly alter the estuaries in which they are constructed (Cleland 1992). Canal construction can destroy parts of wetlands, mangroves and salt marshes, areas that are important as fish nursery and feeding grounds, and as refuges for migratory birds. Water quality can be impaired by excess fertiliser from gardens and lawns, by sewage or septic overflows adding nutrients to the waters, and by organic litter decomposing and reducing dissolved oxygen levels in the canals and estuary. Canals may act as silt traps for sediment from urban runoff, and unless they are oriented along prevailing winds and dredged to suitable depths, they may be poorly flushed by tidal flows. Lowering of the water table may draw down water levels in nearby wetlands and lead to compaction and subsidence of organic sediments in the wetlands. Toxic substances are another potential problem, either as a result of boat use (fuel, litter, anti-fouling agents and sewage if pump-out facilities are not available), or as a result of drainage from acid sulphate soils excavated during canal construction.

Queensland's Gold Coast has the greatest concentration of canal subdivisions in Australia. The canals are excavated, and adjacent land is built up above flood levels, sometimes up to 10 km from the mouth of an estuary (Morton 1992). The channels in the estuary may be modified outside the canal area to provide a single deep channel to replace shallow and shifting channels. Although seagrass beds may survive in the estuary, they do not establish themselves in the canals. Canals located near the estuary mouth may be well flushed, and avoid the problems of poor oxygen levels and high silt loads that characterise canals constructed further inland. However, even if water quality is good, there are adverse effects on fish populations. The canals may have species numbers and fish densities comparable to the environments they replace, but the species mix is different. Small fish of no commercial or recreational value (herrings and silver biddies) occupy the canals because the habitat provides their food stock of plankton in the water and tiny invertebrates on the bed (microbenthic fauna). The fish sought commercially and by recreational anglers feed mainly on vegetative detritus and larger bed fauna (macrobenthic invertebrates). These fish—such as mullet, whiting and bream—are much less common in the canals (Table 6.2).

Table 6.2
Fish populations in canals and coastal wetlands, Gold Coast, Queensland (number of species, and percentage of sampled populations)

Location	*No. of species*	*Percentage of herrings/ biddies*	*Percentage of mullet/whiting/ bream*
Salt marsh	17	0.2	58.0
Mangrove	30	0.5	34.6
Canal in mid-estuary	38	62.9	26.0
Canal at estuary mouth	35	69.7	11.5

Data calculated from Morton 1992

Urbanisation along the coastline has affected not only the estuaries but also the beaches. Blowouts from coastal dunes moving across roads and other property led to demands for stabilisation. Bitou bush was introduced to stabilise drifting sand and has become a noxious weed, infesting 60% of the New South Wales coast. Beach erosion during the mid-1970s was severe along the east-coast urban areas, as storm waves removed the sand from beaches and cut into foredunes that had been unwisely developed. Images of houses collapsing into the sea appeared on television and in newspapers. Local councils were forced to dump huge boulders or concrete blocks to make walls to protect other buildings. This permanently diminished the beaches; as the storms subsided and some sand was washed back onshore, the waves reflecting off the walls carried the sand back out to sea before it could be deposited along the beach area. Demand for restored beaches along the Gold Coast prompted beach-nourishment programs. Sand was dredged from rivers, or mined from coastal dunes, and sent to the eroded beaches. Unfortunately, the programs were often unsuccessful. River sand was too fine to remain on the beach, and was eroded and transported out to sea. Other coastal works have also led to beach erosion: stabilising mobile sand dunes has reduced sand supply to beaches on Phillip Island in Victoria; offshore dredging has increased wave attack, and thus erosion, at Bribie Island near Brisbane; constructing breakwaters at the mouth of the Tweed River has interrupted longshore movement of sand and led to erosion further north at Coolangatta (Bird 1988). Artificial opening of the Murray River mouth in 1981 has been followed by the loss of 45 ha of vegetated dune from the Sir Richard Peninsula and highlighted the need for understanding coastal processes when making management decisions (Harvey 1996).

Unlike the streams in south-eastern Australia, the rivers of north Queensland carry large volumes of sediment to the coast, particularly during floods caused by cyclonic rains. Agricultural and forestry operations in their catchments lead to high erosion rates and supply sediment to the streams. Yet the subtidal delta of the Barron River, near Cairns in north Queensland, has grown very slowly since a major change in channel location took place as a result of a flood in 1939. When the Tinaroo Dam was built on the upper Barron River in 1958, it trapped the sand-sized load and some of the finer load of the river. A smaller weir further downstream was thought to trap sediment only in low flows and to release this sediment by scouring during floods. Coupled with bed erosion below the dams, the net sediment yield of the river should have been maintained. However, the delta did not continue to grow, and sand that once moved northwards from the delta onto beaches was no longer available, so the beaches north of the river eroded. Why was this so? It might have been that the estimates suggesting that there would be no change to sediment yield were wrong. It might have been that dredging of sand and gravel from the lower reaches of the river for construction purposes greatly exceeded the authorised limit. Whatever the precise cause, it seems clear that disturbance of the river's sediment flow is the reason for beach erosion (Pringle 1991).

In Cleveland Bay, north Queensland, seagrass beds almost disappeared at some time between 1961 and 1974 (the dates of available aerial photographs) but then recovered by the mid-1980s. Dredging associated with port development at Townsville could have been the cause of the loss of seagrass, via increased turbidity after dumping of fine dredge spoil. However, two severe cyclones and associated flooding (which led to high turbidity and also freshwater inputs) could have caused or contributed to the decline (Pringle 1996). Again this illustrates the difficulty of separating natural and anthropogenic causes (Box 11).

Box 11
Nutrients and the crown-of-thorns starfish

Large outbreaks of the crown-of-thorns starfish, *Acanthaster planci*, have affected many parts of the Indo-Pacific coral reef systems over the past 20 years. On the Great Barrier Reef questions surrounding the causes and best management of the outbreaks have been highly controversial. The starfish preys on live coral and can strip the coral cover of a reef within a couple of years. The coral may take 20 years to regrow, and re-establishment of full biodiversity of the reef would take even longer. Obviously, an infested reef has little tourist potential, yet management options are limited. Outbreaks on the Reef probably began in the 1950s, but were noticed mainly after the mid-1960s. The starfish spread southwards from Cairns to near Bowen at a rate that varied from about 50 km per year in 1966 to 80 km per year by 1974. Another infestation began near Cairns in 1979, and had spread 150 km northwards and 560 km southwards by 1991.

The starfish spawns in late spring to early summer, at the same time as many of the hard coral species. Clouds of larvae are produced, and these spread the infestation as they are carried by currents. The outbreaks have occurred in areas where there are crescentic or patch reefs. Areas of fringing and submerged reefs are less affected, and lagoonal and ribbon reefs seem the least susceptible. This may be due partly to the reef morphology. Crescentic reefs may act as traps for the larvae. Also, they have a long perimeter in relation to their area, and the coral on which the starfish feeds is found along the reef slope, so there is a relatively abundant food supply. Some researchers argue that the outbreaks are simply natural fluctuations; others see them as evidence of environmental degradation. There is evidence that the starfish has been present on the Reef for thousands of years, but until recently it might not have occurred in such large numbers and so frequently. Higher nutrient levels in the waters of the inner reef lagoon might have increased nanoplankton and chlorophyll levels, and provided abundant food for the starfish larvae. Another theory is that overfishing of the triton shell has reduced predation. There is considerable dispute about how much nutrient levels have changed, and no conclusive answer to the reason for the starfish outbreaks. Interestingly similar infestations of the coral-eating snail *Drupella* have occurred at Ningaloo, again with no clear reason being defined.

The Reef has been subjected to increasing environmental pressure: population growth, extension and intensification of agricultural activity, expansion of tourism, and industrial development have all affected the coastal area, particularly by altering the quality of runoff entering the reef waters. About 15% of land-derived nutrients into reef waters come from cane farms, most of the remainder from natural areas or grazing land (via soil erosion) and <1% from urban areas. Except after storms when sediment is stirred through the water column, phosphorus levels in the water over most of the Reef are not enhanced significantly by these inputs. More and larger ships are using the shipping channels because of the development since the 1960s of aluminium and nickel refineries on the coast, and of export facilities for coal from the Bowen Basin. Eutrophication has affected reefs near the outfalls for sewage at resorts, and the Great Barrier Reef Marine Park Authority has responded by requiring longer outfalls and the introduction of tertiary treatment before discharge. This does not solve the problems about which there is most debate: how widespread are raised nutrient levels due to urban and agricultural runoff, and what are their effects? For example, the cyanophyte *Oscillatoria* can bloom and cause 'red tides' (Plate L). The organism is a nitrogen-fixer and can bloom in nutrient-poor waters. It does not necessarily indicate eutrophic conditions. However, even though it does not require high nitrogen levels, blooming may be encouraged if phosphorus levels are high; and the nitrogen fixed by the organism from the air may then be released into the water, worsening the nutrient levels.

A study near Innisfail showed that both nitrogen and phosphorus from fertilisers moved downstream, associated with sediments but under different physico-chemical conditions. In the estuary and near-shore zone, the levels of both nutrients were very low, despite the high levels in the stream. Phosphorus levels were as low as those on the Great Barrier Reef or in the estuary of a stream with a pristine catchment. It seems that the nutrients were removed from the water by microbial activity, and that pollutants from land are trapped within the reef lagoon and are not transported to the outer waters. Thus, there appears to be a consensus that parts of the reef waters (for example, the Cairns region) are adversely affected by polluted stream flow, but there is little agreement about how much wider the problems may be.

Data from Brodie 1997, Eyre 1993, Hopley 1988, Hopley et al. 1989, Kenchington 1990, Moran et al. 1992, Walker et al. 1991, Zann 1995

Sand mining

Sand is taken from coastal areas for two reasons: as a source of heavy minerals (rutile, ilmenite, monazite and zircon, known together as the 'black sands'), and as construction and foundry material. It is ironic that, of either of these uses, the heavy mineral industry has received the greatest opposition, since this operation does not permanently remove all the sand from the dune. It initially involves the mining of all sand from the dune; but then the required minerals are separated (Plate M), and the unwanted (largely siliceous) sand is replaced as re-formed artificial dunes. These dunes are then revegetated with forest or heath, to reconstruct the former vegetation. The replacement vegetation is not an exact replica of the original forest or heathland. Fauna such as lizards may not return to the sites for at least five years, and even after 15 years the forest is different from undisturbed forest and includes many introduced species (Buckney and Morrison 1992, Twigg and Fox 1991). In the best cases, the program of rehabilitation is considered highly successful, and the Bridge Hill and Jerusalem Creek mining areas of Myall Lakes are now included in national parks. Nevertheless, the controversy over this and other sand mining, and the legislative response to public opposition to the industry, were cited as the reasons why the sand-mining company stopped operating in New South Wales. Their operations moved to Western Australia, where environmental control focused on payment of a bond to ensure that rehabilitation was properly carried out (Roberts 1992). More recently, the mining in eastern Australia has moved on-shore, extracting minerals from relict Tertiary shorelines in the lower Murray Basin.

The earlier move brought an interesting period of conflict between industry and the conservation movement to an end (Morley 1981). Mining of black sands on the New South Wales beaches began in the 1870s, to gain the small reserves of gold, platinum and tin in the deposits. Most mining occurred after storms, when the deposits were exposed and concentrated by erosion and wave action. As early as 1895 the mineral seam was followed inshore into relic beach deposits under the old dunes, and mining began at the Jerusalem Creek site. However, here and at other sites it was only economical to remove the easily accessible material, and activity was spasmodic. The Second World War, and the Korean War in 1950, gave a huge impetus to the industry. The rutile and ilmenite, once thought useless, became important sources of titanium, and zircon became useful in steel-making. By the 1970s New South Wales was the largest world supplier of rutile, and sand mining was a major export industry for New South Wales and Queensland. More reserves were needed, and companies moved into the high old dunes behind the beaches. This led to extensive surface disturbance and clearing of forests in areas that conservationists were seeking to have declared national parks. In 1965 the New South Wales government set up the Sim Committee to try to resolve the dispute between conflicting interests, but there was little progress. In 1972 a company was prevented from mining red gum forest at Wyong. In the same year controversy erupted at Myall Lakes; State and Commonwealth government departments supported a group of conservation movements, led by Milo Dunphy, in opposing the mining of two strips of old high dunes. After an inquiry, permission was given to mine one strip but not the other. In 1977 the State government banned future sand mining in national parks; a vigorous program to place more coastal land under national park protection began soon after; and a policy of preserving coastal wetlands was introduced. These changes led to the withdrawal of the mining companies from the State.

Sand mining also prompted one of the Commonwealth's early interventions in environmental issues. Leases to mine mineral sands on Fraser Island were held under Queensland mining legislation, but the separated concentrates could only be exported with Commonwealth government permission. Conservationists, using the acronym FIDO (Fraser Island Defence Organisation), protested against mining, and the Commonwealth held one of the first inquiries under its *Environmental Protection (Impact of Proposals) Act 1974*. The Minister for Minerals and Energy refused the export licences, and after a subsequent challenge by the company, the High Court ruled that the Minister's decision was valid. He was able to consider any matter, including environmental impact, when deciding whether to grant an export licence (Environmental Defenders Office NSW 1992). This was an example of the way in which the Commonwealth used its powers over other matters (here, its customs power) to influence environmental issues that were otherwise under State control (Fisher 1980). However, this victory in the mid-1970s was by no means the end of controversy over the use of Fraser Island. Proposals for sand mining and logging of its rainforest continued. Fraser Island was more effectively protected when the area was listed under World Heritage legislation.

Mining sand for construction and for glass making involves complete removal of the dune sands. Some of this market is satisfied by river-sand deposits, but the major urban areas along the coasts place high demands on the dune resources as well. For many years the dunes were mined, leaving large pits, which provided convenient garbage disposal sites. In 1974, during storms and severe beach erosion in New South Wales, some of the remaining sand walls between the sea and these pits were breached. Polluted water and debris flowed out to sea, polluting the water and the beaches nearby (Plate N). At that time researchers in south-eastern Australia were beginning to recognise how finite the coastal sand resources were. Until the dunes of south-eastern Australia were dated (using radiocarbon techniques) most experts believed that Australian beaches were like those in California, where large rivers transport a constant supply of sand down to the coast. It was thought that this sand could then flow along the coast. But it was found that even the sand just behind the beach was a few thousand years old and that rivers carried very little sand out to sea. Most of their sediment was very fine, the coarser sand being trapped at the back of coastal lagoons. The sand near the shoreline was blocked by rocky promontories and remained in 'compartments', moving onshore and offshore in response to storms, but not moving along the coast. Thus, the sand locked up in the dunes, on the beaches and in the nearshore zone was a very limited resource that had stopped moving onshore some 3000 years ago. And this sand could not be mined without upsetting the long-term equilibrium. Sand mined from the back of dunes left a 'sink' into which other sand could blow under onshore winds; this sand was lost and could not be moved back shorewards, so beach erosion was likely. Although there were some relic deposits in the back dunes, which could be mined without disturbing the sediment balance on the beaches, often these were environmentally sensitive because of the forest that had developed on them (Young and Reffel 1981). Mining of coastal dunes became an increasingly unacceptable activity, and alternative sources (mainly of river sands) were exploited. This, of course, simply moved the problem from one environment to another, rather than truly solving the problem of demand for a commodity with a high environmental cost.

Opposition to mining of coastal dunes is still strong. In the 1970s the spectacular and silica-rich dunes of Shelburne Bay on Cape York came under threat. Like Fraser Island further south, Shelburne Bay has the unusual feature of rainforest developed on sand dunes, and associated wetlands. A proposal to export over 250 000 tonnes of silica sand was abandoned in 1987 after a storm of protest. The controversy was exacerbated by proposals to establish a space launching station in the region. The Commonwealth government again intervened, as it had in the case of Fraser Island, and refused to give foreign investment approval for the sand-mining operation. The Wilderness Preservation Society has suggested a large integrated system of reserves. These reserves incorporate both existing national parks and parts of land currently under Aboriginal reserve (Horstman 1992).

Conclusion

Although the discovery of Australia by Europeans depended on crossing the oceans, and although its largest settlements have developed around ports focused on international sea transport, Australia's self-image is largely land-based. Australians' concerns with the sea are more coastal- than ocean-based, and it is in the coastal zone that the environmental impacts of European settlement are most obvious. It is not known whether these impacts have caused any extinctions of fauna or flora, although—because they have been far more localised than terrestrial impacts—such extinctions seem unlikely. Only recently has there been a significant effort to set aside marine conservation reserves, and at present these reserves are located overwhelmingly in the Great Barrier Reef Marine Park. Exploitation of marine resources has turned from whaling (banned since 1980) and some fishing, for most of Australia's history, to expanded fishing activity, aquaculture, and development of offshore oil and gas deposits since the Second World War. In line with the general growth of environmental awareness, the environmental consequences of these developments are coming under increased scrutiny. Australians have come to realise that, as with their soils on land, the productivity of their seas is poor in comparison with many other parts of the world; hence, present levels of exploitation of marine biota in Australia may not be sustainable.

However, it is the impacts along the coastline that continue to draw most attention: the possibility of oil spills, the mining of dunes for minerals or for their entire sand content, the environmental deterioration of estuaries, the disruptive outbreaks of natural and imported organisms. Since these are so closely related to environmental pressures resulting from major urban developments, Chapter 7 will explore the environmental impacts of cities.

Chapter 7

Urbanisation and Its Effects

Introduction

WRITING IN 1976 (in the days before inclusive language became mandatory) in a book titled *Man and Landscape in Australia,* the architect Stuart Murray commented that

> The built environment of today could be described as consisting of high-energy, purpose built short life buildings. The first two terms, high-energy and purpose built are readily understandable, high-energy because of their high consumption of materials and energy, purpose built because of the specialist use buildings that the user demands. Short life is the inevitable end of such special purpose buildings. They do not generally adapt to the changing needs of society. The rapidly growing demand for urban buildings and spaces has a tremendous impact on the natural environment. No other activity of man, except perhaps war, literally removes whole blocks of natural landscape as [does] the extraction of building materials.
>
> Murray 1976, p. 240

While experts may decry the quality and planning of our cities, people flock to them. Another contributor to the book just quoted from pointed out the impossibility of decentralising urban development in Australia, and this observation has been borne out by experience. By 1991, 85.3% of Australia's population lived in urban areas (defined as places with populations greater than 1000). People and industries cluster in the capitals and in the major cities of the more populous States; the fastest growing areas are coastal and close to major settlements: the Gold Coast/Tweed Heads (New South Wales/ Queensland) and the Sunshine Coast, Cairns and Hervey Bay (Queensland) were the only statistical areas to grow at 4% or more in 1981–91 (ABS 1994). This means that coastal and urban issues—sewage disposal, catchment clearance, shoreline erosion, estuarine modification—are interlinked.

Most of Australia's capital cities and centres of rapid urban growth are in the most comfortable climatic zones (Figure 7.1). Yet climatic and other natural hazards can affect all of Australia's cities, and large populations mean that impacts are exacerbated. Only the northern cities are likely to be affected by tropical cyclones, but the devastation of Darwin by Cyclone Tracy on Christmas Day 1974 sent shock waves through the entire Australian com-

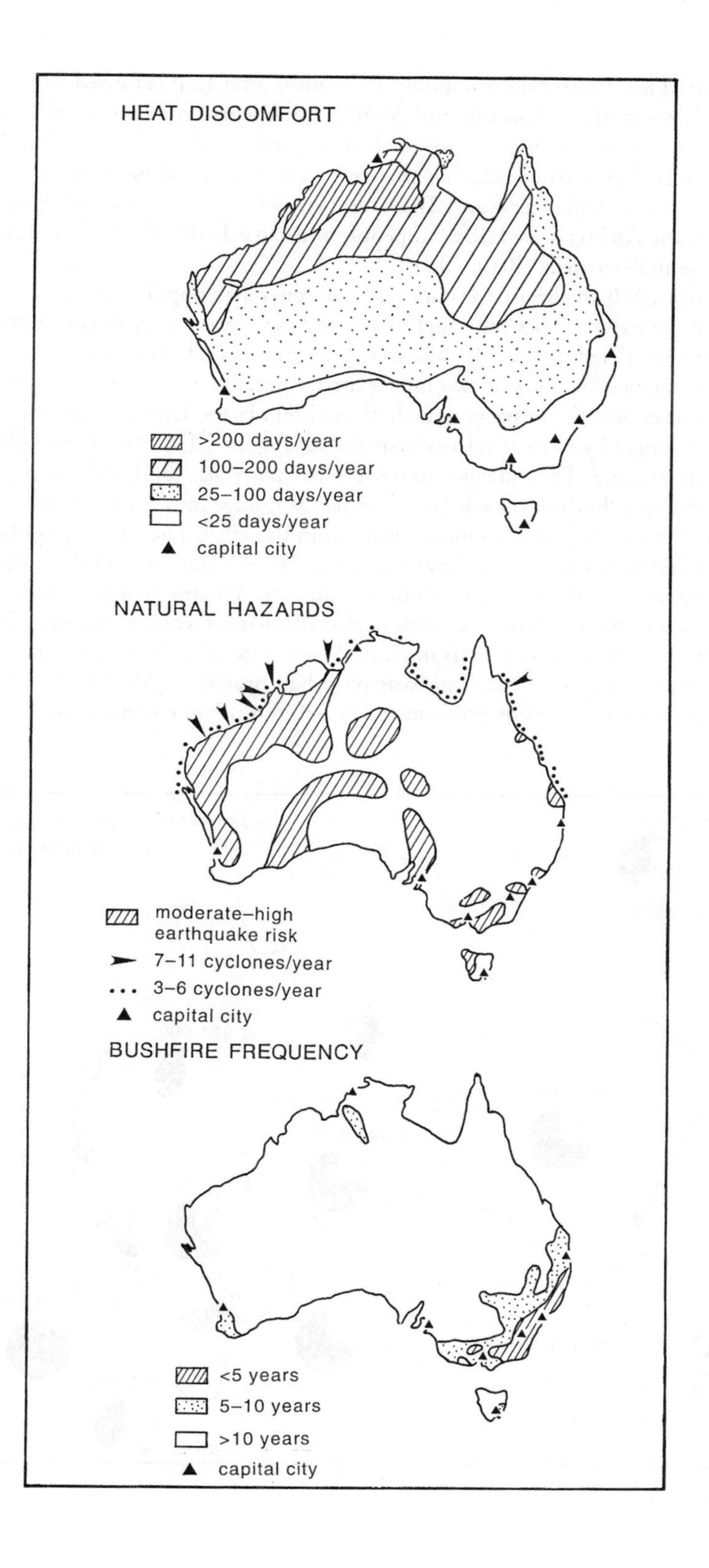

Figure 7.1
The location of capital cities, relative to heat discomfort zones and natural hazard risk. (Source: data from AUSLIG 1992.)

munity. This disaster occurred only 11 months after floods caused by Cyclone Wanda resulted in 35 deaths and A$30 million worth of damage in Brisbane. Most of our major cities are in, or close to, areas in which there is a moderate to high risk of earthquakes, but this fact seemed to be of little interest to the public or to city planners until 28 December 1989, when a quake of magnitude 5.5 on the Richter scale killed 12 people and caused A$1000 million worth of damage in Newcastle (Bryant 1991).

Since 1976 an almost constant 64% of Australia's population has chosen to live in the capital cities, although the percentage varies in different States and Territories (Figure 7.2). To assess the environmental consequences of this choice, we need to look at two interrelated aspects: the impacts within the city boundaries and the impacts outside those boundaries. Even though the actual area occupied by cities is relatively small (see Figure 3.1, p. 34), their influence is wide-ranging. They are net importers of materials, food and people; they export wastes both purposely (for example, as sewage into oceans or rivers) and indirectly (as air pollution moves away from the city). Their large populations mean that urban viewpoints have a major political influence, and this influence impinges on land management outside the city. Yet one writer considers our cities to be, in many ways, our most neglected environments (Kenworthy 1987), while others note an emphasis in Australia on non-urban environmental issues, in contrast to a worldwide emphasis on urban problems (Mercer et al. 1994). What then are the major problems related to our urban environment?

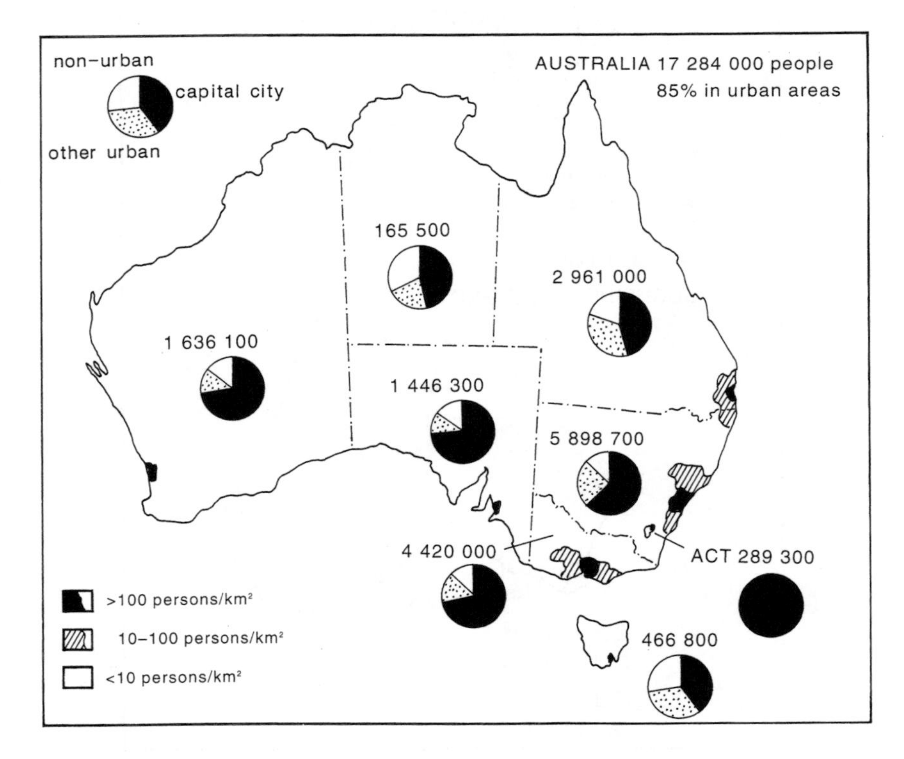

Figure 7.2
The distribution of the urban population in Australia, showing the percentages of each State's or Territory's population in metropolitan and other urban centres.
(Source: data from ABS 1994.)

Defining the problems: What are the issues?

The Australian Conservation Foundation has addressed this question in two issues of its journal, *Habitat*—one titled 'Urban Spaces, People Places' in 1987 and another titled 'Better Cities: Meeting the Challenge' in 1992. Both issues called for more public transport and a reduction in the use of cars; both preferred the infilling existing urban areas with compact and medium-density housing to continuing expansion of suburbia. The more recent issue reflected the current —and increasing—involvement of the Commonwealth government in setting the environmental agenda, whereas the 1987 issue dealt with more pragmatic and day-to-day matters, such as toxic waste disposal, pollution, land-use conflicts, the health effects of powerlines, and creek regeneration. While we can argue that there has been some progress on the pragmatic matters, it is not easy to envisage any significant slowdown in the use of cars or the spread of suburbia. Indeed, we may see the increasing demand for rural residential or hobby farm development as another stage in the expansion of the suburbs. J. Kenworthy (1987) comments that suburbia reflects rural ideals and a desire to escape city life; hobby farms are surely the result of a desire to escape a little further.

Australia State of the Environment 1996 (Department of Environment Sport and Territories 1996, ch. 10, p. 20) identified four areas of concern:

- dispersed and low density population, with the attendant high use of cars and resultant air pollution
- waste management because of high consumption of resources and high levels of waste produced
- remnant vegetation and habitat within cities
- changing patterns of population, leading to social and cultural pressures as well as environmental impacts.

In large cities conflicts over the management of particular sites are common. One study examined reports in two major metropolitan newspapers, and identified 60 conflicts in Melbourne and 71 in Sydney over a six-month period in 1989. In neither city were protests only mounted and participated in by groups of articulate, middle-class people fighting to preserve the high amenity of their own neighbourhoods:

> Locational conflict is endemic in Melbourne and Sydney . . . The individuals likely to contest decisions of government or initiatives in the private sector are likely to be found everywhere . . . Most common overall are concerns about noxious nuisances and transport . . . For the most part, the main conflicts are between, on the one hand, individuals and pressure groups, and on the other hand, government, especially state and local government.
>
> Humphreys and Walmsley 1991, p. 326

Confrontations between government and residents are not surprising. Main roads and railways, water supply and sewage disposal, planning of urban expansion, power supply, toxic waste disposal, and policing of environmental control are all the responsibility of large and powerful statutory authorities, responsible to State government, but effectively largely autonomous in their activities. Yet their actions, combined with the actions of local governments, determine urban environmental quality. However, while conflicts occur across the cities, the pattern is not uniform. A study of resident action groups in Sydney shows that these environmental activists focus on different issues in different places. By their nature, resident action groups are concerned with local, not global, issues, and the local issues are highly variable. About half the groups in long-established suburbs were formal associations monitoring the decisions of their local councils. The other groups in these suburbs were formed to fight decisions made in regard to urban development (about 20% of groups), traffic concerns, facilities/services, and environmental/heritage matters (less than 10% of groups each). In the newer, outer suburbs there were fewer formal associations, and the overwhelming concerns were in relation to urban development and industrial development. These differences reflect the longer histories of groups in the inner suburbs, and also the trend to move controversial developments, such as 'mega-tips', to the city periphery (Costello and Dunn 1994).

Cities as importers

Cities are importers of materials and energy. Sand and gravel come from beach dunes and river floodplains, to be used in concrete for buildings, foundations, kerbing and driveways. Blue metal is quarried and brought in as road base, as a component of some concretes, as railway ballast and as drainage material. Timber is harvested from forests and plantations to provide building frameworks, fencing and furniture. Food is brought in from the hinterland, from other States and from overseas. Very rarely can a city supply its needs from within its boundaries. For example, some sand and gravel for Sydney is still dredged from rivers in the urban area, notably the Georges River, but much comes from the periphery or from outside the city. The extensive Penrith Lakes scheme involved a dramatic land-use change. In the 1970s the river terraces along the Hawkesbury River, west of Penrith, were farmed, mainly for citrus fruit. Now they have been mined to extract the sands and gravels from below the fertile topsoil, leaving vast open pits (Figure 7.3). These will eventually be filled with water and used for urban recreation. It could be argued that the city of Sydney now extends beyond Penrith into the Blue Mountains, but clearly this area of sand extraction lies at least on the city's periphery. Thus, although Australia's mining industry is generally export-oriented and geared to production for the overseas market, there are some mining operations that exist because of domestic demand. The environmental effects of these are indirectly the result of urban development.

The way in which the tentacles of the city reach out is perhaps more obvious if we look at the importing of energy and water. Melbourne's electricity comes mainly from the Latrobe Valley, 70 km to the south-east. Shallow deposits of brown coal are mined by open-cut methods and taken directly by conveyors to power stations designed specifically to take advantage of the characteristics of the coal supplying them. Power generation is so important that it takes priority over other land uses such as grazing. The rivers have been diverted, and old towns moved, to allow access to high quality coal. Dams for cooling water have

Figure 7.3
Pits excavated to supply sand and gravel to the Sydney urban area, Penrith district, New South Wales.

been constructed. Removal of groundwater from the pits has led to ground subsidence over a wide area. Air pollution due to the power stations is monitored by a sophisticated system, which allows adverse effects to be identified, but also allows stacks for emissions to be designed efficiently (Manins 1986). Similarly, large coal-fired power stations supply electricity to most of the large cities (Figure 7.4). Sydney's power comes from the Lithgow area, 80 km west, and the Hunter Valley, 100 km north; Adelaide is supplied from Port Augusta, 250 km north-west, using coal from Leigh Creek, a further 250 km north; Collie power station is 150 km south of Perth.

Figure 7.4
Wallerawang coal-fired power station near Lithgow, New South Wales.

Water is imported into the city, used and modified, and discharged as sewage or runoff. Melbourne draws water from several catchments, including the Thomson River, 100 km east of the city limits. The Sydney supply came initially from a small dam at Prospect, then from the headwaters of the Nepean and Woronora Rivers to the south (Figure 7.5). As the city grew, these sources became inadequate, and the very large Lake Burragorang was impounded behind Warragamba Dam on the Wollondilly River. Still more water was needed, and so the Shoalhaven River was dammed at its junction with the Kangaroo. Future plans envisaged yet another dam on the Shoalhaven—at Welcome Reef, close to Braidwood—some 175 km beyond the present southern limit of the city, although this has been delayed until at least 2030 by strategies to reduce per capita consumption in Sydney. Consumption rose from 388 L/head/day in 1968 to 530 L/hd/d in 1981 but has since fallen to 425 L/hd/d in 1997, due to pricing incentives and water-saving devices (especially single flush toilets). The target is 328 L/hd/d in 2010. As 67% of consumption in 1994/95 was by residential users, changing the behaviour of domestic consumers is crucial (Sydney Water 1995 and pers. comm.).

On a nationwide scale, the amount of water used in cities for domestic and industrial or commercial purposes is significantly less than the volume used in rural areas (Figure 7.6). However, the dams supplying urban areas are large, and the management of their catchments is often a controversial issue. The controversies over forestry and bauxite mining near Perth, for instance, have been fuelled by concerns that these uses of the catchment will cause salinisation of the Perth water supply. But it is not only the effects of commercial activities that are controversial. The Sydney catchments are largely alienated from public use, although there was some political pressure in 1986 to open the dams for recreational uses. This has long been accepted practice on other water supply dams, such Lake Eildon, north-east of Melbourne. However, the move for the main dams near Sydney to be used in this way was defeated, largely because of the support given to the Water Board by conservation groups, who recognised the ecological protection given by the catchment areas remaining closed.

Cities as exporters

While cities import materials and energy, they also export wastes. Again, water is a good example. Disposal of sewage is a major problem for all large urban areas. Each person in Sydney generates about 200 litres of sewage every day (Warner 1991). This is discharged either into the major rivers of the Sydney area (the Georges and Nepean–Hawkesbury) or directly into the ocean. Pollution of popular surfing beaches caused storms of protest, and in 1991 the Sydney Water Board began to use long, deep ocean outfalls. The aim was to take treated sewage away from beaches that were recognised as severely polluted and to encourage greater mixing and dispersion. However, the sewage seems to have dispersed only across the inner shelf (up to 10 km offshore) and to move parallel to the coast, rather than moving well away. A chemical produced in the digestive tracts of animals, coprostanol, was traced as an indicator of sewage, and was found accumulating in the sediments of the inner shelf. Although the

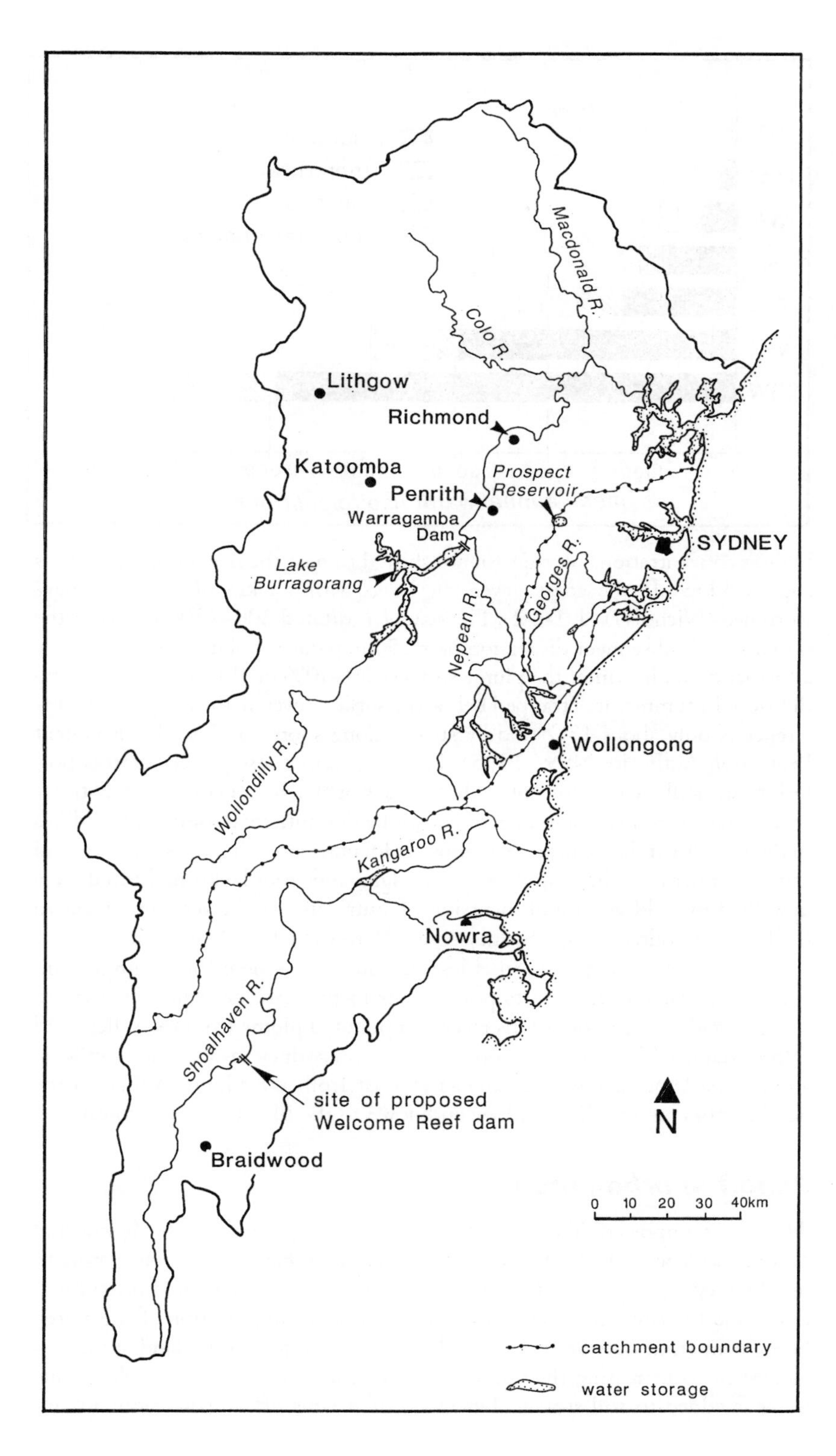

Figure 7.5
The catchments from which Sydney's water supply is drawn. Prospect Reservoir was built by 1888, the dams in the headwaters of the Nepean and Woronora Rivers between 1902 and 1941, Warragamba Dam by 1960, and the Shoalhaven scheme in the 1970s.

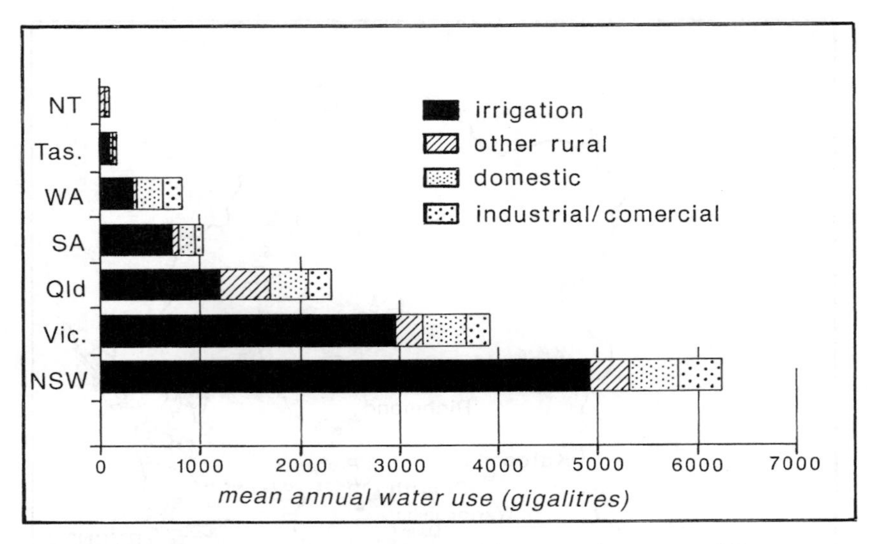

Figure 7.6
Bar graph showing mean annual water use in Australia. (Source: data from ABS 1992.)

highest concentrations of coprostanol shifted to near the new outfall positions (up to 4 km off the coast), even the levels within 1 km of the coast have worsened (Nichols et al. 1992). The sewage is diluted 200–1300 fold near the outfalls, and subsequent dispersion over 15 km reduces pollutant levels only by a further 50%. In winter the plume surfaces 20–40% of the time, encouraging dilution. In summer it is trapped below the surface (even if the temperature difference is only about 1°C) and by strong along-shore currents (Environment Protection Authority NSW 1995). The long-term consequences of this pollution are still being monitored, but it is worth remembering that nutrient enrichment of near-coastal waters may not be due entirely to pollutants. Eddies in the East Australian Current can move cold, nutrient-rich waters onshore, and cause increases in the populations of algae and zooplankton. Indeed, this upwelling of cold water and incursion of nutrients may be very important to the fishery resources of the Sydney region (Cresswell 1994) (see Plate L).

The export of air pollution is less obvious, but is nevertheless important. Melbourne sometimes exports polluted air to the Latrobe Valley, 100 km to the east, and on other occasions receives a polluted plume from the valley (Bell 1989, Manins 1986). Circulation of air around Sydney involves flows moving around the large region that extends at least from the Hunter Valley to the north, across metropolitan Sydney and down to the Illawarra in the south.

Runoff in urban areas

The water imported into the city is not only used within the house and exported as sewage. R. F. Warner (1991) estimated that 60% of water brought into Sydney was used outside houses: for irrigation of gardens and horticultural crops, and for yard and car cleaning. The effect of the importation of this water, brought in from outside the urban districts and added to the land surface, is equivalent to increasing the rainfall of drier parts of the city by 30%. Some water is added to soil storage, but much of it runs off. Furthermore, urbanisation increases the amount of rainfall that runs off the land surface. Hard surfaces such as concrete paths, tiled roofs and asphalted roads do not allow rainfall to percolate down into the soil, and drainage systems are designed to carry runoff rapidly away and prevent infiltration. As Warner notes, 'The urban surface is a major source of environmental contamination, and surface rubbish in all shapes and forms is very efficiently washed into stormwater drains and then into creeks and rivers by the highly organised, enhanced urban runoff' (Warner 1991, p. 5). Less water is lost from urban land than from uncleared land because there are fewer trees and thus lower evapotranspiration. After urbanisation, streams reach higher flood peaks in a shorter time and carry higher discharges than before.

In the 1920s and 1930s there were few cars, and many roads in Sydney and other cities were unsealed and poorly drained (Figure 7.7). Since then, Australian cities 'have been transformed from low car-owning, compact, public transport and walking oriented places into endless, car-oriented suburban tracts' (Kenworthy 1987, p. 3). Sealed surfaces and enclosed drainage systems were constructed to avoid the sorts of problems shown in Figure 7.7. Unfortunately the spread of suburbia has occurred during a period of increased rainfall since 1949. A change to a wet climatic regime has coincided with the spread of the city, the increase in car ownership, and the demand for sealed roads and piped drainage.

Not surprisingly, therefore, urban flooding is still a problem. Continuing subdivision of land up into catchments can feed increasing runoff into old systems, causing overloading. As more and more water is carried by drains at the downstream end of the catchment, the capacity of the pipes is exceeded more often. Design standards for drainage can vary within a catchment, as different local councils set different standards. During floods that occurred in Sydney after intense rainfall in November 1984, velocities along gutters were measured

Figure 7.7
Flooded road in Maroubra, Sydney, about 1925, before roads were sealed, kerbed and guttered to control stormwater. (Source: Mitchell Library.)

at 2–4 m/second. Flooding was most rapid and severe in the upper reaches of large catchments, and where stormwater pipes surcharged because the flow was too great for them to carry it. But natural, open channels carried even these very large flows with little damage resulting (Riley et al. 1986). High rainfalls lead to urban floods. which bring damage, extensive evacuations (particularly in the low-lying Georges and Hawkesbury River floodplains) and sometimes deaths. In August 1986 six people were killed when they were swept away or electrocuted by falling power lines, and 1600 people were evacuated. These storms over cities are, as the Sydney Water Board aptly describes them in an educational video, 'hard rains'. In August 1998 devastating floods hit Wollongong. Newspaper reports called them 'one in 300 year' events but in fact the rains which produced them were intense but localised falls that are characteristic of the Wollongong climate. The damage was overwhelmingly within the urban areas and along watercourses that had been diverted, piped or otherwise modified. It was the urban fabric rather than an exceptional rainfall which made the floods so severe (Plate O).

Alteration to the hydrological regime of an area as a result of urbanisation can have a range of effects other than just increased surface runoff (Warner 1991):

- Weirs and dams can change the pattern of flow, and trap sediment that is moving downstream. They can prevent the upstream migration of fish and of tidal inflow. These changes can significantly alter the ecology of a stream, particularly in the estuarine reaches. Flushing of pollutants and nutrients down the river can be reduced, leading to associated impacts such as algal blooms.
- Channelisation and the concrete lining of drainage ways increase discharges and flow velocities, and therefore increase stream power. This greater stream power encourages erosion, particularly if the flow has a low sediment content.
- Drainage of swampy land can increase the flow downstream.
- During clearing for subdivisions, large volumes of sediment are often washed into urban streams. The sediment is dumped further downstream in unsightly deposits, which may block low water flows and create stagnant pools in the watercourse.
- Eroded sediment washed into estuaries can reduce the volume of the tidal prism, and thus the tidal flushing of the estuary.
- Mining of sand and aggregate from large streams leads to scour of the river bed and to erosion upstream. In the Nepean–Hawkesbury system, bed and bank erosion resulting from mining is accentuated because the water-supply dams in the headwaters trap sand that would otherwise be moving into the mined reaches.

Runoff from urban areas can carry more than the pollutant load of secondarily treated sewage effluent. Levels of pollutants are higher in runoff from established residential areas than from developing suburbs or from rural catchments, except that suspended sediment concentrations are usually highest in runoff from developing areas. High rainfalls mean more runoff and greater pollutant loads, and water quality can be consistently poor as a result. In a recent study, urban runoff entering and flowing through the Botany Wetlands in

Sydney was collected at seven sites on 42 occasions, during both wet and dry weather, over a three year period to July 1992. Its quality was compared with the desirable criteria for urban water (State Pollution Control Commission 1990). Table 7.1 shows the State Pollution Control Commission (SPCC) criteria for some parameters, and the percentage of sampling times that sampled water did not meet these criteria.

Table 7.1
Quality of urban stormwater, Botany, New South Wales

Parameter	*Criterion*	*Exceedance*
Suspended solids	10 mg/L	14–59%
Total phosphorus	0.05 mg/L	56–88%
Total Kjeldahl nitrogen	0.5 mg/L	58–93%
Biological oxygen demand	2 mg/L	18–62%
Faecal coliforms	200 CFU/100 mL	59–87%
Lead	0.02 mg/L	22–71%
Oil	2 mg/L	42–62%
PCB (pesticide residue)	10^{-6} mg/L	not detected

Data from Sim 1993

The catchment above the Botany wetlands is similar to many urban catchments. It is fully developed and only lightly industrialised (about 15% of the catchment is industrial land). It has large areas of open space (48% of the catchment), and the remaining areas are residential, mixed with some commercial uses. About one-third of the catchment is impervious surface (roads, roofs, concreted areas and so on). Thus, the runoff from this catchment is probably typical of runoff from many urban catchments, and its low quality is a matter for concern. Indeed, the poor quality of urban runoff has been a major argument against the continued expansion of the Sydney urban area into large new housing estates in the catchment of the Nepean–Hawkesbury River. The river receives urban runoff and treated sewage, so that, during low flows, over 90% of water in the river is urban effluent. Heavy metal concentrations are close to background levels in the main channel but elevated in several of the urbanised headwaters (Birch et al. 1998).

Waste disposal via septic absorption systems was a feature of many suburbs in Australian cities, and still characterises many small communities. In New South Wales alone there were 250 000 septic systems in 1991; yet studies in both New South Wales and South Australia showed that most systems perform unsatisfactorily. The South Australian study was done in the Mt Lofty catchment, from which 65% of the water supply for metropolitan Adelaide is taken. At one site, below a small village, the watercourse had a total phosphorus level of 1 mg/L, total Kjeldahl nitrogen level of 2.8 mg/L, a coliform level of 14 000 organisms/L and a biological oxygen demand of 6 mg/L. The septic systems responsible were providing waste disposal for small allotments on steep slopes covered by shallow soils with low permeability. The wastes were not retained in the soil for long enough for contaminants to be removed. The septic problem worsens as areas become more affluent because water usage increases with the introduction of automatic washing machines, dishwashers, spas and swimming pools. The systems that are not efficient under low-water usage become hydraulically overloaded under the increase in wastewater discharge and are then even less satisfactory in their performance.

The city and the bush

Cities have grown up on good agricultural land. The ability to grow at least part of their own food supply was crucial to the survival of early settlements, and many other major towns began as service centres to agricultural areas. We should not be surprised, therefore, that the growth of Australian cities has led to the loss of arable land in some of our most fertile regions. Suburban expansion towards the fertile red soils of the Brisbane area was threatening the use of these soils for horticulture by the mid-1970s; their level topography and nearness to the coast has meant that suburbia has now swamped them (Chittleborough 1986). Similar examples are found elsewhere and 1600 km^2 was consumed by suburban development around Australia's five largest cities between 1986 and 1993 (Department of Environment Sport and Territories 1996) (Figure 7.8). Other land is lost by indirect impacts, such as inundation when large dams are constructed to supply cooling water to electricity generating plants whose output goes mainly to the cities.

Studies on the north coast of New South Wales indicated that 95% of land loss in 1970–79 was due to hobby farm development. Only 4% was due to direct urban encroachment and less than 1% to mining developments. The situation is worsened where there is little high quality land available for agriculture. In some regions hobby farming is permitted only on land that is largely unsuitable for cropping. However, continuing demand for rural residential land increases the value of near-urban properties to a level where there is little incen-

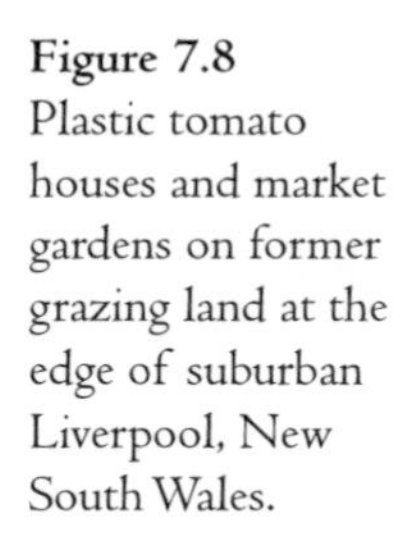

Figure 7.8 Plastic tomato houses and market gardens on former grazing land at the edge of suburban Liverpool, New South Wales.

tive to keep farming, particularly in industries that are time-consuming, such as dairying. In Wingecarribee Shire, to the south of Sydney, land could fetch $10 000 per ha in 1992. Even farmers who wished to continue in production found the financial incentive to sell out of the Illawarra and buy into the North Coast or the western irrigation areas, where land was cheaper, too great to resist. On a statewide basis, this was a disadvantage because of the existing land pressures and problems of the North Coast, and the very great concern about land degradation and salinisation in the semi-arid regions. There is little logic in moving agricultural production to less favourable land when adequate land of poor agricultural quality is available for urban and rural residential use. Nevertheless, the trend to shift agriculture away from urban regions seems to be inexorable and illustrates the considerable indirect environmental influence that the growth of cities can generate.

The problems introduced by the loss of rural land to very small holdings go beyond the loss of agricultural viability. Servicing and environmental costs for roads, water and waste disposal are higher, yet people moving from urban centres to rural residences demand the same services as those to which they have been accustomed. They are often unwilling to accept, as part of their environment, activities that already exist or would normally be acceptable in a traditional rural environment, such as quarries, piggeries or abattoirs. The use of fertilisers or pesticides on adjacent properties may also be seen as unacceptable. And hobby farms contribute to the clearance of our remaining wooded areas:

> The destruction of bush on poor sites has been accelerated by the rural retreat phenomenon. Rural retreaters are urban dwellers with rural fantasies that they attempt to materialise by establishing peasant farms in the bush while living off money gained in the city. Blocks of poor land can be offloaded on to these people, who will develop their properties without the necessity of profit. The irony of rural retreating lies in the destruction of the bush by some of its most romantic devotees.
>
> Kirkpatrick 1994, p. 51

Within cities, few remnants of the original bushland remain, and those that survive are modified. Exotic weeds invade them, especially along creeks where nutrient levels are high (Preston 1995). Native species may expand their range and density. For example, *Pittosporum undulatum* Vent. has become well established outside its former communities in Sydney, Melbourne, Lord Howe Island, Norfolk Island, King Island, Albany and towns in South Australia. In Sydney it has invaded along the edges of suburbia. In older suburbs where fires have been suppressed for long periods, it is well established and has spread deeply into bush reserves (Rose 1997).

The search for living space that combines the comforts of city life with the tranquillity of rural or bushland areas has caused uncertainties about the management of bushland, both within and around cities. In major cities with large areas of bushland reserved in national parks around them, and in places where the suburbs are intermingled with remnant bushland, these uncertainties focus on what may be called 'the appropriate fire'. Devastating fires are an inevitable part of the environment of several major cities—Melbourne and Sydney, in particular. The Ash Wednesday fires of February 1983 burnt out vast areas of agricultural land and killed thousands of livestock, but received the most publicity because they burnt out many houses and were responsible for the death of 71 people (Table 7.2). The effects could have been worse in South Australia if the fires had not been largely downwind of the nearest urban centres—Clare and the Adelaide Hills (Bardsley et al. 1983).

Table 7.2
Losses resulting from the Ash Wednesday fires, February 1983

	Victoria	*South Australia*	*Total*
Deaths (people)	45	26	71
Deaths (livestock)	26 000	328 000	354 000
Houses lost	1 511	196	1 707
Area burnt (ha)	330 000	222 300	552 300

Data from Bardsley et al. 1983

Fuel-reduction burning is unpopular because of concerns about its ecological effects and because of the air pollution it can create. In May 1991 smoke from fires to the south and south-west of Sydney was so dense over the city, particularly the western suburbs, that a newspaper headline read 'Smog is choking our kids!' and there was some evidence of an increase in the incidence of asthma (Churches and Corbett 1991). Restrictions on backyard burning were introduced into Sydney to lessen the 'brown haze' pollution that characterised Sydney in the winter months. After gardening during the day many people lit backyard incinerators on autumn and winter evenings, when temperature inversions, due to rapid cooling of the ground, prevented dispersion and forced the smoke to flow down over adjoining houses. Persistent high pressure systems created higher-level inversions, trapping the smoke in the urban atmosphere for several days. The smoke—combined with dust, sea salt and vehicle fumes—caused poor visibility, and complaints led to the enforcement of a total ban on backyard incineration. All these factors combined to minimise clearing and fuel reduction around the city. However, the catastrophic fires of January 1994 caused a rapid turnaround in government policy, and in public acceptance of fuel-reduction burning, at least in the short term.

Are we choking? Air pollution and its effects

There is a strong public perception that air pollution causes ill health, and worldwide there seems to be a consistent relationship between air pollution levels and respiratory illness. Nevertheless, in Australia, air pollution and bushfire smoke do not trigger statistically significant increases in asthma attendances or admissions. Indeed, asthma levels are higher in some rural than urban areas, but obviously levels of air pollution can exacerbate ill health in people with existing respiratory problems (Australian Medical Association 1997, NSW Health Department 1996). Hence it is prudent to try to reduce air pollution for health as well as aesthetic reasons.

Table 7.3
Ambient air quality standards

Parameter	*Standard*	*Days exceedence*	*Measurement period*
Carbon monoxide	9.0 ppm	1 day per year	8 hours
Nitrogen dioxide	0.12 ppm	1 day per year	1 hour
	0.03 ppm	0 days per year	1 year
Photochemical	0.10 ppm	1 day per year	1 hour
Oxidants as ozone	0.08 ppm	1 day per year	4 hours
Sulfur dioxide	0.08 ppm	1 day per year	1 day
	0.02 ppm	0 days per year	1 year
Lead	0.50 μg/m³	none	1 year
Particles PM10 (particles finer than 10 microns)	50 μg/m³	5 days per year	1 day

Data from <http://www.nepc.gov.au/air right body.html> 24 March 1999

In June 1998 uniform standards for ambient air quality were set for Australia by the National Environment Protection Council (NEPC), a council of ministers from all Australian state, territory and national governments. The standards set the level which could be accepted for a maximum number of days per year, measured over a defined period (Table 7.3). Most of these standards are related to motor vehicle emissions and other major users of fossil fuels. The combustion of fossil fuels at high temperatures—in power stations, motor vehicles and so on—produces nitrogen oxides and hydrocarbons. These emissions are chemically reactive and combine together to form photochemical smog, probably the most widely recognised air pollution problem in the world's cities. The combination of these primary emissions (or precursors) creates the secondary pollutants of photochemical smog, according to the general equation:

nitrogen oxides (NO_x) + volatile organic compounds (VOCs) + sunlight >>ozone (O_3) + re-formed NO_x + other secondary pollutants

(Note that VOCs formerly were called non methane hydrocarbons or NMHCs, and are sometimes called reactive volatile organic compounds or ROCs)

The components of photochemical smog (O_3, NO_x) are oxidants. They are chemically reactive, and ozone is especially corrosive to many materials and is a known health hazard. Currently the Australian standard allows only 10 parts per hundred million (100 parts per billion) as an average one-hour reading for ozone. Until 1998 the level was higher, 120 pphm, and levels in Melbourne and Sydney exceeded this from five to six times per year in the late 1980s. This was a significant improvement on levels in the 1970s, when the limit was exceeded dozens of times per year in Sydney; the worst was in 1976–77 when the limit was exceeded 40 times. The situation was initially brought under control through reduction in emissions of VOCs, by better control of static sources such as petrochemical storage areas, and by fitting catalytic converters to cars. This last measure was possible only after unleaded petrol was introduced in 1986. Because VOC concentrations control the rate of smog formation, lowering those concentrations allowed more time for pollutants to disperse and resulted in less smog over the city. Unfortunately, the flow of air across Sydney means that these dispersed pollutants are not carried away from the city, but move westwards into areas where most of the city's future population growth is expected. This means that authorities now must face the more difficult task of controlling the emissions of nitrogen oxides, since the initial concentrations of these determine the maximum possible ozone levels. Yet nitrogen oxide emissions have been rising, and potentially very high levels of photochemical smog have been predicted for the western suburbs of Sydney in the future (Carras and Johnson 1983, Wright 1991–92).

Specific meteorological conditions cause photochemical smog to build up to high levels in Sydney (Figure 7.9) (Carras and Johnson 1983, Wright 1991–92). The precursors (vehicle emissions) are emitted in the morning and blown westwards across the suburbs by sea breezes during the day. When there is a high pressure cell over the region, skies are clear and ultraviolet radiation is high, providing energy to convert the emissions to smog. The smog cannot escape to high altitudes, because a high pressure cell creates a subsidence inversion that traps the air at low levels. In the night cool air drainage (katabatic winds), flowing down from the plateaux surrounding Sydney, drift the polluted air back towards the inner city. Some of the pollution moves out to sea and then blows back onshore under the next day's sea breeze to affect the southern suburbs, and sometimes Wollongong. Some of the polluted air simply picks up

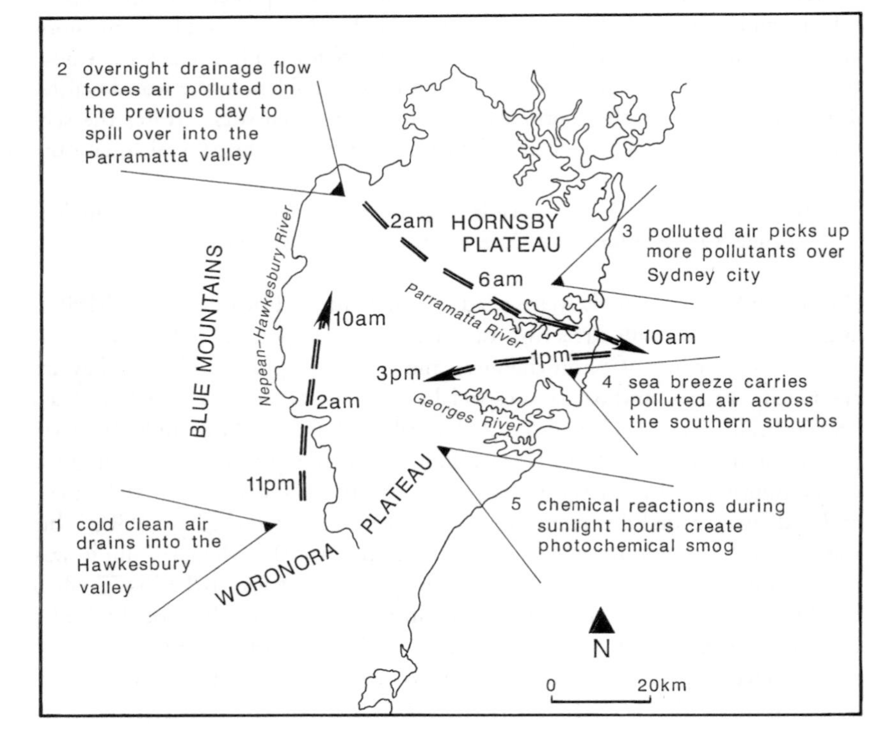

Figure 7.9
Circulation of polluted air and the build-up of photochemical smog in the Sydney region. (Source: data from Carras and Johnson 1983, and Wright 1991–92.)

more pollutants over the city next morning, and then even dirtier air is moved westwards through the day, so that smog and ozone levels increase. Until a front moves through, there is little relief.

In Melbourne there is no ring of plateaux surrounding the city to trap the air flow that blows onshore with the sea breeze and to generate a night-time flow seawards over the city. Also, the sea breeze directs some of the polluted air away from the city, not back across it. However, on days with a low level inversion and weak easterly winds, and also under the influence of land- and sea-breezes, a clockwise eddy can recirculate polluted air over Melbourne, leading to high ozone levels. Polluted air swirls around Brisbane, Perth and Adelaide under similar conditions, but in these cities the ozone guideline is exceeded only occasionally (Department of Environment Sport and Territories 1996).

In Melbourne, Sydney, Brisbane and Perth, over the last 10 years, median anthropogenic levels of ozone have been steady. Better emission controls have been offset by more vehicles, and, in all cities on 20–40 days per year, there is an underlying loading of ozone of 4–6 pphm. As this is about half the exceedence limit, it is clear that, as population grows and vehicle use grows even more rapidly, significant increases in air pollution can be expected in the next decade (Box 12). This is so especially as new emission controls target VOCs and CO (carbon monoxide) but may do little to reduce NO_x (Reid 1997).

On the subject of ozone, one further point needs to be emphasised. The production of ozone at low levels in the atmosphere is an adverse impact of the emission of pollutants. Ozone in the air we breathe is a health hazard. However, ozone at high altitudes, well above the level of the air we breathe, is a natural occurrence and necessary for our continued health. It absorbs excessive ultraviolet radiation and lowers the risk of diseases such as melanomas on the skin and cataracts in the eyes. Hence, there is concern about the damage to the natural ozone layer that appears to be caused by gases such as CFCs, and Australia is a signatory to a United Nations agreement to control the emission of these gases.

In the urban regions of the Northern Hemisphere, acid rain is a recognised problem of international significance. Nitrogen and sulphur oxides released by burning fossil fuels cause acidity, not only near their source areas, but downwind for long distances across country borders. The acidification of lakes in Scandinavia is blamed largely on emissions in Great Britain and western Europe. Australian urban agglomerations are much smaller than their Northern Hemisphere counterparts, and acid rain is not a major issue here. Locally, over a large city such as Sydney, and probably also Melbourne, rainfall is acidified. Upwind of Sydney, pH is typically above 5, but downwind, it is typically between 4 and 4.5. Hydrogen ion (H^+) concentrations are higher in Sydney rainfall than in rainfall in the industrialised Hunter and Latrobe Valleys, where power station emissions cause comparably high sulphate and nitrate levels. However, the levels of all ions are two or more times lower than values found even in rural areas of the Northern Hemisphere. For the most part, rainwater acidity in Australia is not significantly greater than one would expect it to be as a result of natural fluctuations (Bridgeman 1989) and even the large power stations of the Latrobe Valley do not add significantly to acid deposition (Holper 1995–96).

Box 12
Motor vehicles and their impacts on city air

Automotive emissions are responsible for many air pollutants: more than 70% of nitrogen oxides, 90% of lead (in 1988, although this declined rapidly after unleaded fuel became widely used), as well as significant amounts of greenhouse gases, carbon monoxide, sulphur oxides, and benzene and other hydrocarbons. Lead levels in fine particulates collected in the Sydney region in 1992 correlated well with motor vehicle density. Even the industrial areas of Newcastle and Wollongong had lower levels of fine particulate lead than inner Sydney. Isolines of lead levels paralleled the major roads connecting the urban centres, and a bulge of relatively high levels south-west of Sydney indicated flows of polluted air into this region of low elevation (Figure 7.10). In Perth, where all petrol is supplied by the Kwinana refinery, it was possible to correlate the atmospheric lead levels with the lead added to all petrol types at the refinery. The lead added varies between grades of petrol, and within grades depending on the type of crude oil that is being refined. Hence, the total lead added at the refinery is the best indicator of the total lead emissions from motor vehicles, and it correlated closely with average lead levels in Perth's air between 1982 and 1987. Not surprisingly, therefore, the introduction of unleaded petrol in 1986 has had a significant effect on lead levels in cities but sale by volume of unleaded petrol only exceeded that of leaded petrol in 1995.

Traffic-flow patterns affect the emission of pollutants. As people drive from the suburbs into the central business district (CBD) of a city, they stop more often and for longer; and they need to accelerate more often, but their average speed is lower. The ratio of carbon monoxide (CO) to nitrogen oxides (NO_x) increases, but this enhances formation of photochemical smog. Vehicle numbers are greatest close to the CBD, and driving conditions there cause higher emissions, especially of hydrocarbons and carbon monoxide. However, this is not an argument for more freeways, and thus for driving styles more like those of the suburbs, in the CBD; providing better roads usually just increases traffic and negates any improvement in emissions as a result of easier travelling. About 60% of travel in Perth's CBD is by private car, and encouraging more use of public transport is a better solution to traffic congestion than more road development. A combined strategy by the National Road Transport Commission and the NEPC to improve vehicle design standards, fuels, brake noise and maintenance, and to reduce traffic congestion, was announced in 1999.

Data from ABS 1992, Cohen et al. 1994, Lyons et al. 1990, National Environment Protection Council 1999a, O'Connor et al. 1990, Reid 1997

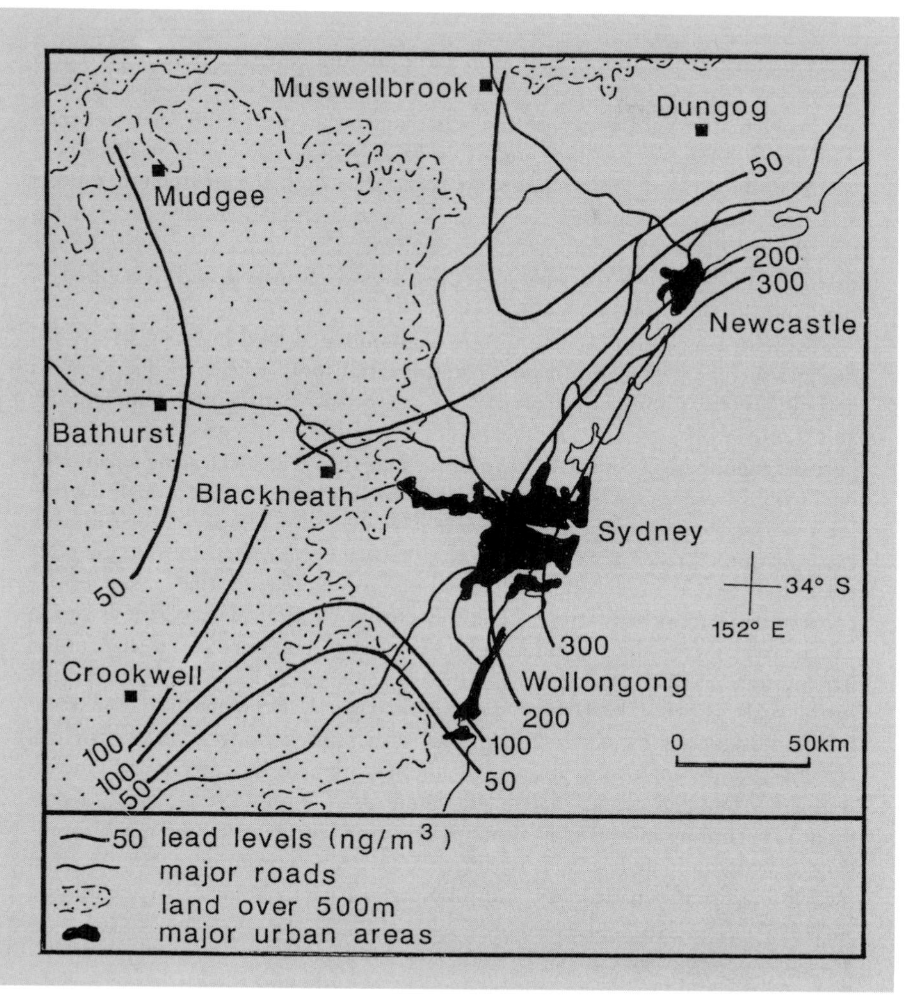

Figure 7.10 Lead levels in fine particulate air pollution around Sydney in 1992, showing the close association of lead pollution with urban areas and major roads. (Source: redrawn from Cohen et al. 1994.)

Solid wastes and contaminated land

The greater the population, the more waste is produced. Of all waste disposed of by councils in Australia, 71% comes from households and 20% from the retail and commercial premises that serve those households. Only 7% is generated by industry, but of course industry disposes of much of its own waste rather than sending it to council facilities. The remaining 2% is produced by government establishments, and again they may dispose of some of their own waste.

Where can all this waste be put? Old quarries, the swales of sand dunes, areas of bushland and other places generally out of sight have been the usual options. In major cities the public health risk of rotting garbage is avoided by covering it with soil or waste such as coal washery spoil; in country areas, the tip is often left burning and a plume of smoke can be traced to the site. The water that leaches through landfill can be highly polluted, and leachate from many older tips can contaminate streams. Significant quantities of nitrogen, biological oxygen demand, chromium, lead, iron and cadmium enter the Botany Wetlands in Sydney from an old refuse site (Sim 1993). Future uses of waste disposal sites are problematic. Subsidence of the waste occurs as putrescible material rots. This process releases methane and other gases, which may combust and which often produce unwelcome odours. At some sites now, the methane is tapped and used to generate power. Accumulated metals and other substances mean that the land is too contaminated for occupancy, even after subsidence of the material ceases. Often the sites are covered with clean soil and developed as parkland or golf courses.

Waste disposal is only one cause of contaminated land in cities. Heavy industries such as steel works or power stations, transport centres such as airports and railway yards, mines or quarries, and chemical or petroleum manufacturing and storage works are often established on the outskirts of cities, but eventually become engulfed by suburbia. The industry may then relocate, leaving a contaminated site whose former use may be forgotten over time, so that an inappropriate use—such as residential development—may be made of it. As suburbia expands onto former agricultural land, potential contamination must be considered. A study near Liverpool, west of Sydney, found that many horticulturalists and chicken farmers kept no records of chemical usage, yet as late as 1961 saturation of the ground with DDT was recommended as a treatment for lice infestations in chicken sheds and the pesticide was still being recommended in the early 1980s. The lack of data resulted partly from the fact that many horticulturalists were recent immigrants with limited knowledge of English, but was due mainly to the absence of any requirement by government, and to frequent name and brand changes in chemicals (Bergin 1997). Local government is becoming increasingly aware of this problem, but making a full inventory of contaminated sites is a difficult and time-consuming task. Written records are usually inadequate, and inventories have to be made by techniques such as:

- searches of business directories and telephone books
- searches of old planning scheme maps to identify land zoned as industrial
- interpretation of aerial photographs to identify small factories, petrol stations, quarries and landfill depots within residential areas, and market gardens or poultry sheds in areas that could become urbanised
- cross-checking of these sources to exclude duplicate sites and
- consultation with long-term residents or local government employees to confirm suspected sites or to specify the precise use of a suspected site.

Soils in urban areas are also contaminated by fallout of air pollution, especially near industries such as smelters, and near roads and other transport routes. Studies in Adelaide showed that lead and zinc levels in soils were considerably higher in the urban area than in the nearby rural Mt Lofty Ranges (Table 7.4). The concentrations of metals were analysed in two different ways: as total levels—by completely breaking down (chemically digesting) the soil so that all metal was released—and as EDTA levels—by extracting the metals in a solvent, EDTA, to release the metals that are probably bioavailable and able to be taken up by plants. The EDTA levels are usually lower than total levels because some metals are tightly bound up in the crystal lattices of clays in the soils and cannot be extracted by the solvent. However, the EDTA method used was exhaustive, so most of the metals in the soils were analysed even by this method. There is little agreement about how best to analyse metals in soils so as to estimate the real risk to health or ecology. For example, some researchers suggest that lead should be extracted in concentrated hydrochloric acid, as this mimics human stomach acidity and indicates how much ingested lead could be expected to be absorbed by the body. The depth at which soil is sampled, and the method of sampling, can also affect the values determined. Nevertheless, comparisons using a consistent technique—as in those reported in Table 7. 4—show clear differences between urban and rural soils in levels of lead, zinc and, to a lesser extent, copper (Tiller 1992).

Table 7.4
Metal contamination of soils in the Adelaide region (median values in mg/kg)

Metal	*Urban area (total)*	*Urban area (EDTA)*	*Rural area (total)*	*Rural area (EDTA)*
Lead (Pb)	50.5	49.0	18.0	8.8
Zinc (Zn)	58.0	24.0	18.0	4.1
Copper (Cu)	15.0	10.0	12.0	2.4
Cadmium (Cd)	0.27	0.15	–	0.16

Data from Tiller 1992

How then do we set an acceptable level for soil contaminants? How do we assess when an urban soil is too contaminated for its existing or proposed use? Guidelines established overseas, particularly in the Northern Hemisphere, were used initially, but it is now recognised that these are not suitable for Australian conditions. Australia's geological history differs from these countries significantly. It is tectonically stable, and only very small areas were affected by Pleistocene glaciations. Australia does not have the young unweathered soils, and the extensive shallow deposits of recent coarse sediments, that characterise much of Europe and North America. Our soils are weathered and nutrient-deficient; they have been affected by aridity over very long periods; they often have hard crusts (laterite, bauxite, calcrete and silcrete) near their surfaces. Hence, they are poorly permeable and naturally acidic, and they may have naturally high accumulations of metals. Consequently, our groundwater aquifers are deep, saline and often relatively high in metals and other contaminants. Whereas over 50% of domestic water is supplied from groundwater in the USA and the Netherlands, only 15% of Australia's domestic supply (and only in Perth, among the capital cities) is from groundwater (Mulvey 1992). Thus, overseas concern about contaminated land often is related to concern for groundwater quality; in Australia concern is more with surface disturbance and consequent air pollution, with plant uptake of contaminants, and with leachate entering streams.

To better understand the difficulties, let us look at the example of aluminium. Because of the high clay contents of our soils, this is an abundant element;

levels may naturally reach 10 000 mg/kg (equivalent to parts per million, or ppm). Yet it can be toxic to plants and to aquatic life at low concentrations in water (less than 5 ppm). Its toxicity depends on the form it takes. When bound chemically (chelated) with organic molecules, it is not toxic, but as the free monomeric ion, it is. At low pH (less than 4.5), aluminium can be mobilised and released from soil into solution; if the organic content of the soil is very low, much of the aluminium may remain as the free ion. These conditions of low pH and low organic content are common in Australian soils, so we can expect toxic levels (particularly for exotic plants unadapted to Australian soils) to be naturally quite common and easily exacerbated by contamination. Yet aluminium was not listed in the 1992 Australian and New Zealand Environment and Conservation Council (ANZECC) guidelines for investigation levels for soil contaminants, and the pH range given in that list (pH 6–8) was atypical of natural soil pH in most major cities (Mulvey 1992).

Thus, there are many problems for governments grappling with the management of contaminated land:

- continuing and increasing demand for landfill sites, despite efforts to divert waste by recycling strategies
- difficulties in identifying sites contaminated by past activities
- rezoning of contaminated sites, such as former petrol stations. Fast food outlets or houses may replace petrol stations without the change being flagged during the planning approval stage
- variability in methods of collection and analysis, and hence in relating different data sets to one another to gain a comprehensive picture
- multiple pathways for contaminants to enter plants, animals and people, and variations in the toxicity of different forms of contaminants
- uncertainty about the effects of different contaminants on human health because of the complexity of possible effects, the difficulties in isolating the effects of a contaminant from other possible causes of similar symptoms, and the possible synergistic effects of several contaminants
- community suspicion of 'scientific' answers to many of these problems.

In March 1999 the Commonwealth government released a draft National Environment Protection Measure (NEPM) and Impact Statement for the assessment of site contamination. The draft NEPM noted that there were at least 80 000—and probably 200 000—contaminated sites, and that the current cost of site assessment is about $60 million per year. The NEPM aims to review and gain acceptance for the ANZECC/NHMRC (National Health and Medical Research Council) guidelines, to establish standard procedures for assessment and to achieve enactment of mirror legislation to provide for nationally consistent assessment (National Environment Protection Council 1999b).

Drawing the battle lines: High-voltage power and health

Environmental controversies in cities are often intense. Population and political influence are centred in the city, where media coverage is swift and extensive. Many people are affected by environmental change; yet this change may be brought about by the demand for an increasingly high quality of life. Local, single issue controversies may trigger significant regional or statewide policy changes. A good example is the controversy over construction of high voltage power lines to bring an assured supply of electricity into cities. In Melbourne the proposal to construct a line from Brunswick to Richmond caused so much protest that both the Victorian and New South Wales governments set up investigative reviews (Gibbs 1991, Powerline Review Panel 1989). The argument began with concern about the environmental quality of waterways that would be crossed by the powerline, but the possibility that exposure to electromagnetic radiation could cause cancer dominated both inquiries. It was an argument about possibilities rather than scientific evidence:

> The Health Department and SECV [State Electricity Commission of Victoria]—no matter how assured they may feel in their interpretation of the scientific/epidemiological data—cannot realistically expect to allay community concerns by defensiveness or resorting only to the cover of professional science. Conversely, groups such as the Powerline Action Group (PAG), which at times exhibits the same degree of certainty and self-assuredness, cannot expect to dictate the parameters of public policy.
>
> Powerline Review Panel 1989, p. 3–4

> As might have been expected, popular feeling on this question has been influenced, not directly by scientific writings on the subject, but rather by accounts given in the media and in books and articles of a quasi-scientific kind intended for readers who are not scientists . . . Nevertheless, the popular apprehension on this subject cannot be dismissed as idle fantasy.
>
> Gibbs 1991, p. 33

The point at issue was a possible link between exposure to the low frequency electrical and magnetic fields generated by high voltage lines, and increased risk of cancer such as leukaemia and brain cancer. Scientific evidence shows no such link. The World Health Organisation and the NH&MRC agree that people can be continuously exposed to magnetic fields of up to 1000 milliGauss (mG) without adverse health effects. Directly below a 500 kilovolt (kV) powerline the field may reach about 180 mG, but only 50 m away the field reduces to less than 10 mG. Electric shavers, hair dryers and electric blankets may create fields of 10–1000 mG, and household background levels may be 1–10 mG. A major review in the USA concluded that there was no plausible biological mechanism to explain why magnetic fields should cause cancer, and little good evidence that they do (G. Davis 1993). In Victoria there was no common ground: the electricity authority adopted the NH&MRC guideline of 1000 mG; the Powerline Action Group adopted a level of 0.5 mG. In the end, however, both the State government inquiries and the United States report acknowledged that governments had to act as if there could be some link, and that community concern itself is a valid reason for 'a policy of prudent avoidance', even if the scientific evidence is at best inconclusive.

This is one of many examples of conflicts in which 'expert' and 'community' perceptions of the risk of environmental damage are vastly different. An environmental impact statement for a major development is usually the end product of a proposal, not a starting point for decisions on the best option for a development's site, form or appropriateness to an area. While the development proposal may meet all objective criteria, such as zoning regulations or pollution emission standards, the public may feel the decision is imposed on them, with little prospect for change, and may seek reasons to justify their objections:

> The conflict between conclusions based on 'fact' and on values is often not acknowledged . . . The predominantly technical nature of the EIS process and the rules of evidence of legal appeal processes tend to focus . . . disputes . . . on scientific issues. The real dispute, however concerns value choices . . . The health aspects of the powerlines dispute were brought into a debate which focussed initially on aesthetic considerations, possibly because health was perceived to be a more legitimate cause for objection.
>
> Ewan et al. 1993, p. 71

Conclusion

Cities occupy a small proportion of Australia's land area, but exert a strong influence on the entire country. Within the cities, high population density means inevitable environmental pressure. Waste disposal; pollutant emissions; demand for goods and services; demand for space for work, recreation and accommodation; concentration of people in travel routes—all these issues have environmental consequences. Increasingly, the general public's response to these consequences leads to demands for a wide range of intangible impacts, as well as the more traditional issues of measurable pollution and degradation, to be considered. Cities are the focus for this changing set of demands because they are the gathering points for most of our population. Hence, their influence extends well beyond city boundaries, not just in demand for physical supplies such as food and water, but also in management of other parts of the landscape. The strong trend towards greater environmental responsibility has important implications for cleaning up the cities, but also for promoting conservation of natural areas, and the sustainability of non-urban land uses.

Chapter 8

Water Supply and Use

Introduction

Australia's limited resources are fought for by competing interests. Both demand and the intensity of debate about water resources may rise and fall not only with changing economic and social pressures but also with natural fluctuations, especially drought. The examples of the development in the Hunter Valley and of uranium mining near Kakadu, given in Chapter 2, illustrate this point. In the Hunter Valley competing users pay very different prices per megalitre of water because of differing social commitment to their activities and differing acceptance of the desirability of subsidies for users (Day 1996). At 1995 prices, a power station paid $7.60/ML for water from regulated flows whereas irrigators paid $3.65/ML. Neither paid for water taken from unregulated streams or from flood 'surplus' flows. In contrast, residential drinking water was valued at $804/ML and bottled 'spring water' at a massive $2.8 million/ML. At Kakadu, ecology rather than water supply is the issue. Magela Creek is a small stream with limited importance as a water resource. Because of the uranium mining in its catchment, however, its water quality and ecology have been the subject of highly funded and extensive research (for example, Brennan et al. 1992). Moreover, due to the long half-life (8 ¥ 10^4 years) of thorium^{230}, the residual parent isotope in the mine tailings, this research has had an unusually long-term perspective. The site and the structures of the tailing impoundments need to be stable for tens of thousands of years in order to prevent leakage of contaminated water to the surroundings (East 1986).

Assessing need: Where from and how much?

Despite—or perhaps because of—the aridity of much of the continent, an assured and unlimited supply of water has been a goal of most governments in Australia, to 'drought proof' farmers and provide unrestricted supplies for city dwellers. To achieve this goal, dam sites have been identified and protected for decades before development; water has been collected in humid areas and then piped long distances to drier regions. There have been bizarre proposals such as one to haul icebergs from Antarctica and melt them here. There is therefore a strong political dimension to water management decisions. Indeed, Chenoweth

(1998) has argued that the Great Divide in Victoria is not only a physical catchment divide but also a political one.

In the 1960s, Melbourne had periods of severe water restrictions. Its water storage was about 11% of Sydney's, and equivalent to only one year's need (in comparison to seven years for Sydney). Two proposals were put forward to increase storage. The first and most expensive was to divert flow from the Thomson River in Gippsland; the second was to bring water from the Big River across the Great Divide. On the eve of the 1964 State election, the then Premier, Sir Henry Bolte, promised to stop Melbourne from gaining access to the Big River water. The announcement apparently was to secure support from irrigation interests in northern Victoria. The policy continued, even though no new dam was built for nearly two decades. In 1983 the Thomson reservoir was completed, and the divide between Melbourne supply and irrigation supply in the north of the State remains. Since the 1980s, increased demand has been met by further development of the Yarra catchment and the re-use of treated water (Chenoweth 1998). There has been a trend towards more efficient use rather than simply continuing dam construction, a trend mirrored elsewhere in Australia, in remote areas as well as in cities (Smith 1998).

The towns of the Pilbara region of Western Australia grew with the development of iron ore mining (see Box 8). To lure workers to the region, attractive salary packages were provided, and these often included free water supply. Initially water came from major aquifers, but supply was supplemented by dams and a barrage across the Fortescue River. To develop more storage in the 1980s would have been expensive. Also, the drawdown of the Millstream aquifer (supplying the towns of the western Pilbara) had harmed wetlands in the Millstream–Chichester National Park. The mining companies reduced the subsidies for water, or encouraged employees to buy homes and thus become responsible for their own water costs. Public education programs encouraged residents to reduce watering of gardens. In Wickham average annual household water use fell from over 1200 kL in 1984–85 to below 800 kL in 1990, but falls in other towns were less dramatic (Vellacott 1992). Average household use in the Pilbara and also in Darwin has been much higher than in the metropolitan cities. In 1993–94, this ranged from 260–270 kL in Sydney, Melbourne and Adelaide, to 570 kL in Hobart and 700 kL in Darwin. In all places, watering the garden accounted for over 30% of use (Department of Environment Sport and Territories 1996). In the dry seasons of northern Australia, the amount needed to sustain exotic flora in gardens is very high, and encouraging people to grow endemic or at least native plants is a major step towards water conservation. In Perth, households use on average 58% of water inside and 42% outside. Strategies with existing technology for reducing consumption in the area supplied from Perth (which includes the city and also the gold-field towns centred on Kalgoorlie) could cut the projected rise in demand to 2010 by over 11%. The strategies include efficient toilets and shower heads, water-wise gardening and efficient irrigation as the main contributors. Because of Perth's groundwater resources, more private domestic bores for garden supply is the next most important strategy (Smith 1998).

Inter-basin transfers

Inter-basin transfers are an important part of Australia's water management systems. The largest of these are for hydroelectricity generation and water supply (Table 8.1). They involve the construction of large dams, with all the consequent environmental impacts; they have effects also on the diverted and receiving stream channels. The streams from which water is diverted lose much of their natural flow, with consequent impacts on their ecology. In one case, tourism is a powerful ameliorating influence. Water is released over the now usually dry Barron Falls to provide views from the tourist train up to Kuranda from Cairns.

Table 8.1
Inter-basin transfers of water

State	*Transfer from basin to basin*	*Amount (GL/yr)*	*Purpose*
New South Wales	Snowy–Murray/ Murrumbidgee	1130	hydroelectricity
Tasmania	Mersey–Forth	830	hydroelectricity
Tasmania	Derwent–Tamar	698	hydroelectricity
South Australia	Murray–SA Gulf area	350	water supply
New South Wales	Shoalhaven–Sydney area	284	water supply
Queensland	Barron–Mitchell	188	irrigation
Victoria	Thomson–Yarra	140	water supply
Western Australia	Murray–Swan coast	105	water supply

Data from Crabb 1997, Table 17

Flows into the receiving streams may alter the channels significantly. The Sydney water supply system involves inter-basin transfer from the Shoalhaven catchment into the Nepean catchment, mainly as a drought-proofing strategy. Water is pumped from the Shoalhaven up to a dam which has drowned most of the Wingecarribee Swamp. From there it can be sent northwards to the Nepean storage, or down the Wingecarribee River to Lake Burragorang. Trial releases down the Wingecarribee River showed that sustained releases at close to bank-full capacity of the river had the greatest potential to cause bank erosion and channel cutoffs. Short releases of flows large enough to cause minor floods could cause erosion but the channel would have time to recover and banks could revegetate between floods. Long-lasting flows near bankfull capacity, however, transport more sediment, saturate the banks and thus reduce soil strength, and exert continuing stress on the banks and bed (Erskine et al. 1995). The long-term impacts are illustrated in a small scheme in northern New South Wales. Water diverted from the Nymboida River passes through the Nymboida power station and is released into three small creeks of an adjacent catchment. Since 1924, the width of the creeks has increased by a factor of more than 2.5, and bed lowering has exceeded 2.5 m in many places (Broderick and Outhet 1998).

The Snowy Scheme is our largest inter-basin transfer (Figure 8.1). Water from the Snowy River is held in Lake Eucumbene and Lake Jindabyne, water from the upper Murrumbidgee also goes to Lake Eucumbene via Tantangara Reservoir, and flow from the Tooma River (a tributary of the Murray) is stored at Tumut Pond Reservoir. Thus, flow from all three catchments is diverted through the Tumut and Blowering power stations into the Murrumbidgee River. Flow from the Snowy River via Island Bend Pondage and the Murray catchment via Geehi Reservoir goes through the Murray power stations into the Murray River. Almost the entire flow of the Snowy River has been diverted across and through the Great Divide into the Murray and Murrumbidgee headwaters. In severe droughts there is no flow into the Snowy out of the impoundment of Lake Jindabyne. Even flow from below Lake Jindabyne is diverted, with flow from the Mowamba River being sent by aqueduct back to Jindabyne. Hence, the Snowy's channel has become smaller and overgrown with exotics, the saline wedge at the mouth has progressed upstream, and the breeding of Australian bass seems to have been impaired (Environment Protection Authority NSW 1997). There is now pressure to release environmental flows—flows sufficient to support the ecology of the rivers downstream—from the Snowy Scheme dams and from dams on many other rivers. In general the concept of environmental flows involves a 'transparent' dam wall. During heavy rain, some flow is let out of the dam and downstream untouched by irrigators or other

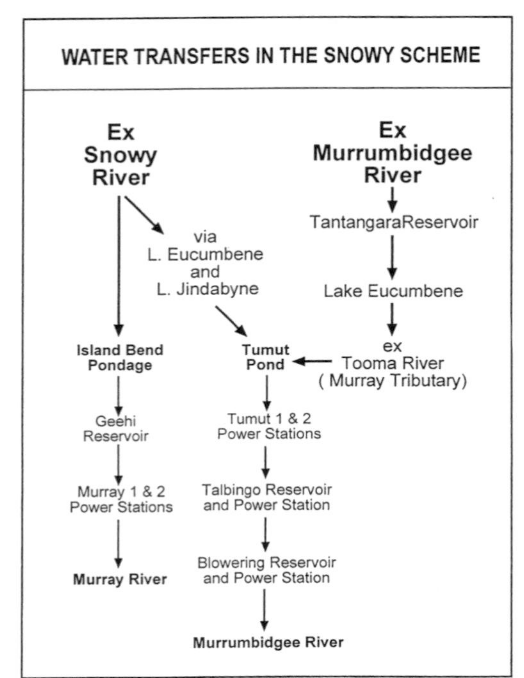

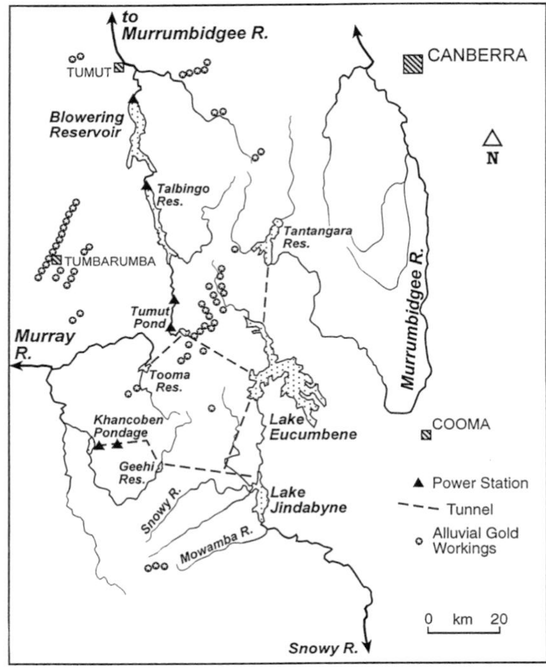

Figure 8.1
The Snowy Mountains area, showing the interconnection of water storages, the hydroelectricity generating stations, and the extent of former alluvial gold workings (Source: Snowy Mountains Authority pamphlets, and 1;250 000 Geology of Kosciusko National Park sheet, 1990).

water users (such as power generation companies, or mines wishing to release and dilute stored saline discharges). When the catchment is dry, less water is released, so flow downstream mimics the natural flow to some extent.

In the case of the Snowy, the matter is particulary complex. Commonwealth, Victoria and New South Wales governments are all involved; conservationists argue that at least 25% of natural flow must be restored; irrigators respond that they have already been hit by reduced water reliability and higher prices, despite being far more efficient in water use than in the past. More than a year after the Snowy Water Inquiry ended, no agreement was in sight (Institute of Australian Geographers 1999).

Impacts of dams and river regulation

There are 447 large dams (walls higher than 10 m) in Australia. Eighty-eight per cent have been built since 1950 (see Figure 2.3) and most of them are for irrigation and/or power generation (Crabb 1997, Smith 1998). Since the 1980s, large dam construction across Australia has declined; this change has been driven mainly by environmental considerations, although it is true also that suitable dam sites are now scarce. Most but not all of our large dams are in areas of humid climate in eastern Australia. The Menindee Lakes Scheme in western New South Wales uses relict lakes as storage areas for flood flows from the Darling River, giving the nearby Broken Hill an assured water supply and providing for irrigation of orchards, grapes and cotton.

Water storage has converted many rivers to a series of dams and weirs. Their longitudinal profile is stepped, their flow patterns and gradients are altered, and seasonality of flow often is reversed. The Nepean River at Penrith, on the western outskirts of Sydney, has been transformed beyond recognition from the bucolic stream painted by John Lewin in 1815, and even the gorge sections further upstream have been largely tamed since Joseph Lycett painted the wild Warragamba rapids (c. 1819–24) and William Piguenit produced his marvellously romantic 'Upper Nepean' of 1888 (Speirs 1981). Its course is now broken into small lakes impounded behind five major dams and thirteen smaller weirs. The impact of the four headwater dams built between 1907 and 1936 to augment Sydney's water supply had a relatively small long-term impact because they blocked off only 6% of the catchment upstream from Penrith. As they were built during the dry climatic phase prior to the 1940s they did cause a noticeable short-term decrease in stream flow. However, the Warragamba Dam, which was completed in 1960 and which blocked off 83% of the catchment, had a major impact on the flow of water and especially of sediment. The immediate effect on flow was offset by the general increase in rainfall after the 1940s, but the long-term impacts have been profound (Warner 1983). The numerous small weirs, built to provide local water supplies, have also changed the river, for now over 93% of the 85 km of channel between Douglas Park and the Penrith weir consists of impounded reaches. Deposition of sand behind the weirs resulted in a loss of channel capacity along this section of the river that triggered extensive bank erosion, and had to be countered by dredging. The main impact, however, is in the reaches starved of sediment downstream from Penrith, where until recently the sediment supply was further decreased by extensive commercial dredging of sand and gravel from the channel. Air photos taken over a period of 40 years along the 42 km stretch from Richmond to Sackville showed that large sections of the banks had retreated by 10–15 m and, in some places, by more than 20 m. And, as considered in the chapter on urbanisation, the degradation of the Nepean–Hawkesbury catchment has been compounded by increasing sewage discharge, so that this is certainly a river under stress (Warner 1983, 1991). But public awareness is focused not on these immediate and serious problems but rather on hypothetical increases in maximum storm runoff that will supposedly accompany greenhouse effects, and there has been a vociferous campaign to raise the wall of Warragamba Dam.

Impoundments themselves have direct environmental effects (Crabb 1997, MacKinnon and Herbert 1996, Smith 1998):

- Dams are like lakes and, in much of Australia, are stratified at least during summer. The upper waters warm up; the deeper waters remain cold and become de-oxygenated. Metals in the bottom sediments may be re-dissolved, leading to high levels of iron, manganese and heavy metals. If water is taken off for domestic use, or released downstream from these deeper waters, its quality may be poor for human consumption, and its temperature and quality may adversely affect aquatic life.
- Shallow margins around dams may provide breeding grounds for mosquitoes, and enhance the spread of insect-borne diseases such as Ross River or Barmah Forest viral fever.
- Water is released from dams at rates and times very different to the natural supply of the streams. Flood flows are smoothed out and the natural fluctuations from very low to very high river levels are replaced by much lower amplitude variations. Often water is stored in winter and released in summer to irrigate crops during the hot and/or dry period of the year. This reverses the natural seasonality of flow.
- The number of minor floods below dams is reduced. Most native fish depend on minor floods to trigger breeding, and the interruption to this has favoured the spread of exotic fish, especially European carp in the Murray–Darling system. Farmers suffer because minor floods replenish both the soil and the soil water storage of floodplains without causing erosion.
- Areas upstream of dams and weirs become permanently inundated. Along the Murray River, areas of river red gums have declined because the wetlands they occupy are now permanently flooded rather than being inundated intermittently.

A significant but often unrecognised consequence of large dam construction is a rise in recreational opportunities for water skiing, fishing and swimming. The small village of Mulwala was transformed after the construction of the Yarrawonga Weir on the Murray River into a resort town with apartments and facilities reminiscent of coastal resorts. Water skiers dodge between the dead river red gums (Figure 8.2). Sand dunes along the Murray west of Albury are the sites for country clubs taking advantage of available irrigation water and of

Figure 8.2
The dead trunks of river red gums mark the drowned Murray River floodplain, inundated to form Lake Mulwala, behind Yarrawonga Weir.

the willingness of golfers to travel. In the alpine areas, trout fishing is a major tourist drawcard.

Although large dam construction has become unfashionable, ironically, large on-farm impoundments have become far more common, especially in the upper Darling catchment (Box 13). Total on-farm storage rose in the Murray–Darling Basin from about 500 GL in 1988 to over 900 GL in 1994 (Crabb 1997). The dams are built to hold allocated water and to take advantage of the right to pump out of the rivers during flood flows. They are shallow and obviously subject to high evaporative losses. In the larger ones, trees are left standing dead in the dam to break up wind-driven seiche effects and reduce wall erosion.

Box 13
Cotton growing in the Namoi and Gwydir valleys

Until construction of the Keepit Dam on the Namoi River in 1960, dryland grazing and wheat growing dominated the agriculture of the Namoi Valley. Once the dam provided an assured water supply, irrigated cotton farming spread rapidly from an initial site near Wee Waa. After the Copeton Dam was built on the Gwydir in 1976, cotton moved into the Gwydir Valley also. Within 20 years, agriculture in the valleys had been transformed. The pasture cover survives along the travelling stock route, but most of the land is cropped under irrigation. Long drains carry water from the river to the fields, sometimes via huge turkey nest dams. It is siphoned onto the fields into furrows between the rows of cotton. Large centre pivot systems irrigate other fields, drawing on the groundwater reserves (Plate P).

The area is underlain by Paleozoic rocks of the Lachlan Fold Belt upstream of Narrabri, and by Jurassic/Cretaceous strata of the Great Artesian Basin downstream. Since the mid-Miocene, about 15 million years ago, 70–120 m of alluvial sediment has accumulated over the basin strata, and the surface topography now has very low relief. The current channel of the Namoi is incised into the floodplain but in major floods water spreads broadly over the area, filling not only billabongs of the current channel but also remnants of much wider channels cut in past more humid climates. The paleochannel of the Namoi was confined in the fold belt strata but built a now-buried broad alluvial fan as it left them. The groundwater used for irrigation comes from sediments of this alluvial fan. This swings north-west from Narrabri, whereas the present course of the river flows westwards (Figure 8.3). The irrigation is taking place in parts of the valley where average annual rainfall averages only about 600 mm, and corresponding evaporation is about 1500 mm, so that there is a deficit of nearly 1000 mm/yr. The cotton industry is dependent on stored surface water and groundwater for irrigation. Only towards the upper catchment, where evaporation is lower and approximately equal to rainfall can dryland cropping of cotton succeed. The Namoi is now the most developed aquifer system in New South Wales and there are embargoes on new licences as well as allocations on existing ones.

Water quality in the lower Namoi is poor. The two highest median levels of phosphorus measured in New South Wales in 1996 were at Mollee on the Namoi (365 mg/L) and Waminda on Pian Creek (275 mg/L). Phosphorus levels in rivers are considered good if <20 mg/L and poor if >50 mg/L at periods of reasonable flow. When the inland steams are at low flow, levels need to be lower to avoid problems of eutrophication. Some of the phosphorus comes from weathering of basalt in the upper catchment but piggeries, cattle feedlots and sewage plants are important sources. Contamination of the water by chemicals used in cotton production is a problem. Local people have complained of illness due to drift from aerial spraying of insecticides and

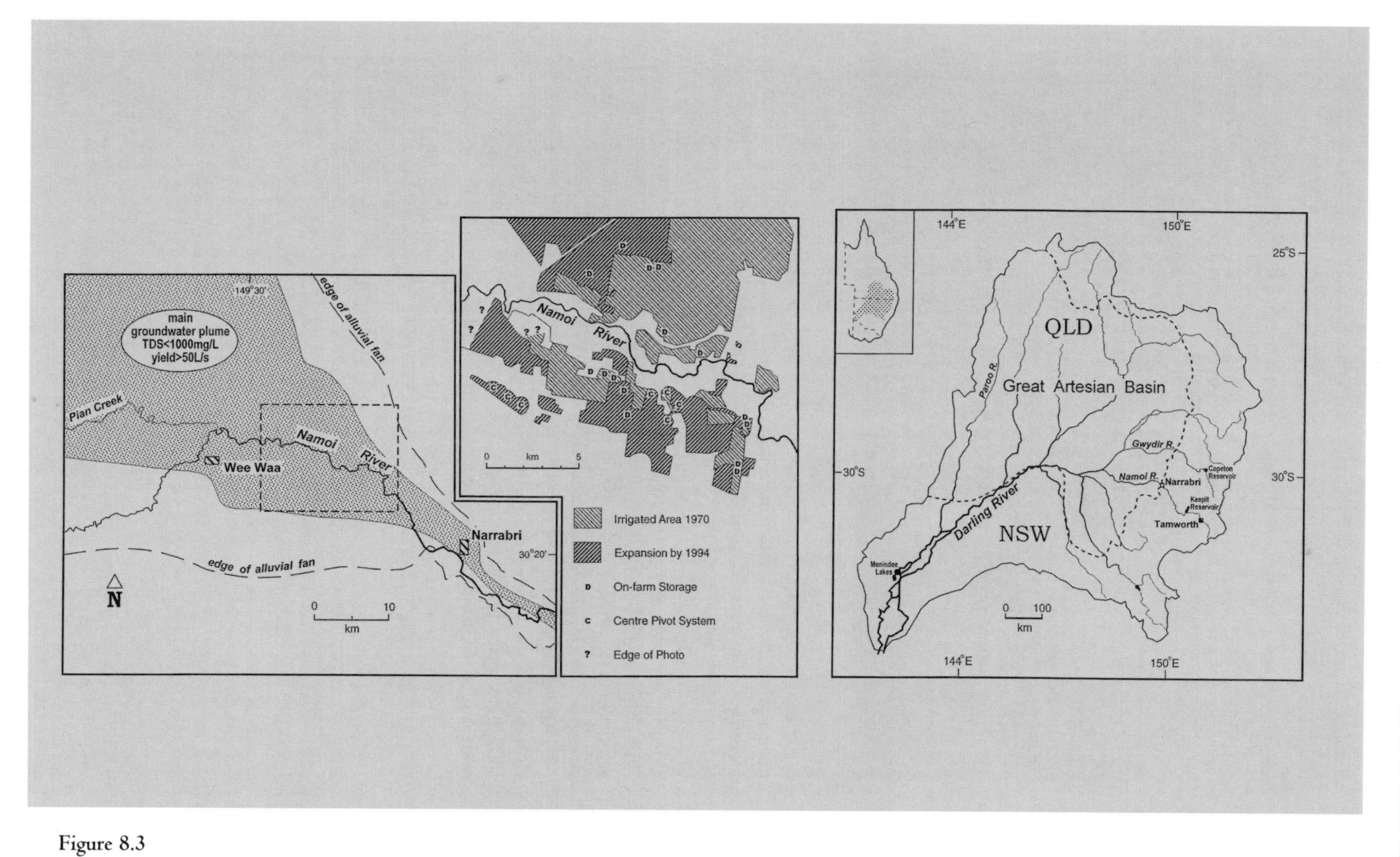

Figure 8.3
Irrigated cotton farming occupying land above the high quality groundwater reserves, Namoi Valley, New South Wales. (Source: author's interpretation of aerial photographs, and the Narrabri 1:250 000 Hydrogeological Series 55–12 Sheet 1988.)

Figure 8.4
Rows of cotton ready for picking, and pipes in irrigation ditch, near Pallamallawa

defoliants. Chemicals used can leak to groundwater, and a 1992 study found 3 of 71 boreholes contaminated with atrazine, a herbicide used to control weeds in irrigation channels. The insecticide, endosulfan, is used widely to control *Heliothus* moth infestations. Unlike other organochlorides, it persists in the soil for only 3–6 months. However, it is acutely toxic to aquatic organisms, and though not proven to be a carcinogen, is harmful to people. Residues from cotton waste used as cattle feed in drought led to beef exports being rejected. Endosulfan use is now being reviewed nationally. Cotton with genetically engineered resistance to *Heliothus* is being grown and, within cotton fields, strips of vegetation to encourage natural moth predators are being trialled.

The wetlands of the Gwydir have declined by about 70% in area because of clearing and dam construction in the catchment. They are still important for birds, and major breeding seasons followed the release of environmental flows and natural floods in the summers of 1995–96 and 1996–97.

Crabb 1997, Department of Environment Sport and Territories 1996, Environment Protection Authority (NSW) 1997, Pigram 1986, Whyte and Conlon 1990

The regulation and modification of rivers has had profound effects also on the river channels and their ecology. Like many rivers in southern and eastern Australia, the Latrobe River in Victoria has been altered in many ways, both deliberately and indirectly by land use changes in the catchment. Three dams have been built in its catchment; there are high temperature and surge discharges from the Yallourn power station. Reinfelds et al. (1995) investigated but rejected claims that there was excessive sediment in the river bed due to sluicing of overburden from the brown coal mines into the river. However, they did conclude that, since the late 1800s, the lower Latrobe River has become wider and deeper, and its banks have become less stable, because of works designed to lessen flooding. The channel has been modified by:

- removal of snags from the river, and clearing of vegetation from the banks, since the 1890s
- construction of drains through levees to allow water to flow off the floodplain more rapidly
- building of local levees to stop creeks breaking their banks
- artificial cutoffs of meanders, which have reduced the overall length of the river by 25% since the 1870s
- dredging to widen the channel.

Ironically, since the 1980s, much of the work on the river has been done to reverse former modifications. Artificial cutoffs have been reinstated, and sections of the banks have been fenced and planted.

However, the impacts of river channel works are overshadowed by those due to water storage, but reversing policies in order to ameliorate impacts obviously threatens the industries based on assured water supplies. Irrigators in New South Wales have argued that environmental flow requirements will jeopardise the economic viability of existing crop systems, especially of rice. Water extractions in the Murray–Darling Basin were capped at 1993–94 levels by agreement of the States and Commonwealth in the Murray–Darling Basin Ministerial Council, primarily in recognition that the resources were highly developed and the environmental problems such as salinisation meant that no further development should occur. Irrigators thus had seen their future entitlements capped or restricted, and opposed further loss of entitlement for environmental flows (Dick 1998). Yet there is continuing pressure for increased irrigation, despite many doubts about the economic benefit and environmental impacts. In 1995 a proposal to establish irrigation on the Paroo River in southern Queensland was strongly opposed and blocked by pastoralists and conservationists. They argued that the Paroo, the only remaining unregulated river in the Murray–Darling Basin, required its natural flood regime to sustain both existing pastoral use and the wetlands in its catchment (Jenkins 1997).

In New South Wales, much of the controversy over environmental flow arose during the prolonged period of drought in the early 1990s. There was concern over the lack of water in major wetlands like the Macquarie Marshes on the Macquarie River, and Cumbung Swamp on the Lachlan River. It was fuelled also by extensive blooms of blue-green algae, not only on the inland rivers but on the Nepean–Hawkesbury system near Sydney.

Algal blooms and phosphorus pollution

In 1991 a severe bloom of cyanobacteria (blue-green algae) affected about 1000 km of the Barwon–Darling River. The blooms covered the surface of the streams in thick slicks and scums, turned the water green, caused taste and odour problems, and released toxins with the potential to cause skin irritations and illness to people, stock and wildlife. The bloom did not develop from one end of the river, but occurred spontaneously over most of its length. Low flows from August to December were a major factor, preventing flushing of developing blooms and allowing suspended solids to settle. The water was clear enough for light to penetrate; it was high in both nitrogen and phosphorus; and it was warm (25–30°C). Flows were low mainly because of drought, but other human-induced factors affected the river. The river at low flow is a series of weir-pools rather than a continuously flowing stream, and irrigation extraction on tributaries further reduced flow to the river. Floods in mid-December cleared the blooms but they had begun to decline before then. The reasons for the bloom and its decline are not fully understood. Certainly the conditions were favourable, but clearing of macrophyte aquatic flora due to European carp infestations in the river may have been influential also (Bowling and Baker 1996). The release of environmental flows is designed to reduce the risk of such blooms in the future.

The other strategy to reduce the risk is aimed at reducing the supply of nutrients, especially phosphorus, to the inland rivers. Phosphorus enters the rivers from point sources, mainly sewage treatment plants and cattle feedlots, and non-point sources, mainly from runoff carrying fertiliser. Control of non-point sources is difficult, and data suggests that point sources are the most significant contributors to the problem (Bond 1998). Effluent irrigation, rather than direct discharge of treated effluent, is a potential solution (CSIRO 1996). Sewage typically contains about 4–10 mg/L of phosphorus and 10–30 mg/L of nitrogen. When this effluent is irrigated onto tree plantations, the phosphorus is used by the trees or is held in the soil. This means it is not easily leached and available to pollute streams. When the trees are harvested, the phosphorus they contain is exported off site and removed from the system. However, the soil phosphorus level usually increases irreversibly, because not all added nutrient is extracted by the growing crop. This may affect future land use (Bond 1998).

Effluent irrigation requires careful planning and management if it is to be sustainable. Firstly, trees or other crops cannot be irrigated during rainfall, or the effluent is lost as runoff and will cause stream pollution. There must be enough on-site storage to hold effluent through prolonged times of wet weather. Secondly, the trees do not use water and nutrients evenly throughout their life cycle. They use water and nitrogen at a high rate for the first few years until the tree canopy closes. After this, their water needs are lower, and some of their nitrogen needs are supplied from decayed leaves on the ground. Hence, the amount of effluent that can be irrigated onto the plantation declines over time (CSIRO 1996). It is important not to over-water, because seepage of excess water into the ground may cause an underlying saline water table to rise. Any irrigation water, and especially effluent, contains some dissolved salts. Total dissolved solids in sewage typically range from 200 to 3000 mg/L (Bond 1998). These are not used by trees or crops, and accumulate in the soil. If the soil of the plantation is not to become too saline, then enough water must be used to flush these salts out of the soil—but not so much that water can seep deep into the landscape! Note, however, that Bond (1998) comments that flushing salt to groundwater may be of little significance if the groundwater is already naturally saline. Furthermore, over time the soil's capacity to absorb phosphorus and cations such as calcium which are not completely removed during plant growth may be exhausted. If this occurs, then any nutrients added in irrigated effluent cannot be stored even temporarily in the soil. Unless they are taken up by the crop, they will move with the soil water and are likely to pollute nearby streams.

Wetlands

Changes to the catchment and river regulation have caused significant declines in many wetlands. They have contracted in area; their ecosystems are less complex and diverse; and the number and abundance of water birds have declined. Even where they are protected by legislation, changes in the catchments may affect them, or the protection offered may be inadequate. For example, in New South Wales, State policy protects coastal wetlands, yet does not cover immediately adjacent freshwater wetlands (Young et al. 1996).

While many of the larger wetlands are recognised as being of international importance and protected under the 1985 Ramsar Convention or migratory bird agreements, over 90% of wetlands in the Murray–Darling Basin are on private property. Not only is this a problem in terms of control but also the status of the wetlands is unknown. Rapid assessment procedures offer a way forward in this respect. Spencer et al. (1998) suggest that the wetlands can be appraised by rating, on a 5-point scale, a range of parameters and then summing the total to provide a wetland condition index. The indicator parameters are bank stability, soil pugging by livestock, soil organic matter content, the width and continuity and height diversity of fringing vegetation, cover and heterogeneity of aquatic vegetation, presence of attached algae and frequency of algal blooms, and the turbidity, conductivity and colour of the water.

Water storage structures usually have little value as replacements for lost natural wetlands. They are usually too deep to sustain aquatic vegetation, and fluctuations in levels mean that their shores are often bare of vegetation which could act as bird refuge. In some cases, however, they may have ecological value. The Wakool–Tullakool evaporation basins west of Deniliquin in New South Wales cover 2100 ha. Since 1988 they have received saline water pumped from below 20 000 ha of irrigated rice fields, a procedure that has lowered the water table significantly. The basins are designed for harvesting of salt and so are divided into bays that have differing depths and salinities. These provide a range of habitats with food and reliable water all year, and security from shooters. They have a rich and at times abundant bird population, including some rare and vulnerable Australian species and 18 species of migratory waders. The Wakool–Tullakool basins are the largest of the 90 evaporation basins in the

Murray–Darling Basin (Roberts 1995), and smaller basins cannot be expected to have the same ecological value. Nevertheless, this example demonstrates that constructed wetlands can be important in increasing the biodiversity within a region that has been degraded by land use change.

Other constructed features can compensate to some extent for lost wetlands. In the coastal valleys near Wollongong, floodplain wetlands have been largely obliterated or greatly altered by clearing and pasture development. Farm dams, although small individually, together constitute as large an area as the single remaining natural floodplain wetland. They occur not on the floodplains but across creeks draining the steep ridges, and their ecology has not been studied. Nevertheless, they are generally shallow and, with other constructed wetlands, add to the depleted wetland stock of the region (Young et al. 1996). Of more obvious ecological value are the reconstructed wetlands in areas previously mined for mineral sands but rehabilitated after mining, in both Western Australia and New South Wales, some of which are now national parkland.

Groundwater

Much of Australia's groundwater is saline but the reasonably fresh resources have been important in the development not only of the arid centre (Box 14) but also the better-watered southern coastal regions. As Figure 2.2 shows, groundwater resources are about 10% or less of divertible surface resources in all the drainage divisions except the most arid (Bulloo–Bancannia, Lake Eyre, Western Plateau and Indian Ocean) and those with Mediterranean climate of wet winter and hot dry summers (South-West Coast and South Australian Gulf). Yet, among those divisions, 30% or more of the groundwater resource is extracted except in the Western Plateau and Indian Ocean divisions, where neither pastoralism nor farming is significant because of the poor soils and arid climate. In the wetter drainage divisions, groundwater is still important, except in Tasmania and the tropical north (Timor Sea and Gulf of Carpentaria divisions). In coastal eastern Australia and the Murray–Darling Basin, again about 30% is extracted. The most developed of these groundwater resources is in the Namoi Valley. This is an example of a region where changing resource assessment has led to significant land use changes; and also of a region where resources are threatened both by over-use and by pollution (see Box 13).

In south-eastern South Australia, good quality groundwater from aquifers in Tertiary sediments, mainly the Oligocene–Miocene Mt Gambier Limestone, has supported the State's most productive agricultural area. The karst topography in the Limestone means that there is no ordered surface drainage network, even though average annual rainfall ranges from 500 to 800 mm/yr across the region. Until the 1970s, the karstic features were seen as ideal disposal systems. Toilet and household wastewaters, effluent from a cheese factory, and carcases of sheep burnt in a bushfire all were disposed of into sinkholes. This led to pollution of the groundwater which was the water supply not only for agriculture but also for towns. These practices were stopped after legislation in 1976 (Emmett and Telfer 1994).

Box 14
Water supply in Alice Springs

Groundwater is essential to the development of the Northern Territory. The following figures illustrate how scarce modern water supplies are, in comparison to other parts of Australia:

Drainage division	*Average precipitation*	*Groundwater*	*Runoff*	*Evaporation*
Western plateau	261 mm	<1%	<1%	99%
South-east coast	909 mm	<1%	17%	82%
Tasmania	1352 mm	<1%	57%	4%

There are substantial volumes of fossil groundwater in many aquifers in the Northern Territory but many resources are too saline for use. Good water resources are often in the sedimentary basins that have poor sandy soil, whereas the water below the best grazing lands is of poor quality. In terms of sustainability, only the Georgina area north-east of Tennant Creek and the Great Artesian Basin intersecting the south-east of the Territory have major divertible resources that are recharged under modern climatic and topographic conditions. Some aquifers hold sufficient water of reasonable quality to allow even irrigation, and an experimental site at Ti Tree yielded good results. However, marketing problems and lack of infrastructure have prevented significant development. In the arid southern region, rainfall averages 200–400 mm/yr but pan evaporation is 3600–4000 mm/yr. Streams flow only for short periods, usually after monsoonal rains.

The Todd River at Alice Springs does not flow for most of the time, but has an average maximum flow of about 150 ML/km^2 in January. Compare this with the Franklin River in Tasmania which has a median flow of 100–250 ML/km^2 throughout the year and maximum flows averaging 150 to nearly 700 ML/km^2. Nevertheless, there was sufficient water stored in the sediments of the Todd River by floods to provide a water supply for Alice Springs from its beginning in 1871 until the 1960s. As the city grew, more water was needed and the Roe Creek borefield was developed. The water is considered to be of good quality, with TDS (total dissolved solids) of 350–600 mg/L. (Note that the maximum acceptable level for Sydney water is 150 mS/m, equivalent to 500 mg/L.) Much of the water is used for gardens and lawns. There is evidence of increasing salinisation of the supply. More importantly, studies have shown that the borefield is not replenished by regional flow but only by seepage from local streams. The aquifer is being mined for water, and new resources will soon have to be developed. Future strategies must include curbing demand and re-using treated wastewater.

Department of Environment Sport and Territories 1996, Jolly and Chin 1992, McDonald 1992, Smith 1998

Water quality concerns have been prominent in other regions where groundwater has been important for agricultural development and where urban expansion has placed increasing pressures on those resources. In the Piccadilly Valley north of Adelaide, land use traditionally has been rural—market gardening, orchards, vineyards, grazing. There is now an increasing residential population and an expanding industry bottling mineral water. Sampling in 1994 indicated that nitrate and bacterial levels could become of concern, but that pesticide levels were lower than in surface waters (AGSO 1998). In Perth, groundwater has been used both for agriculture and for garden watering. It supplies about 60% of Perth's water requirements (Department of Environment Sport and Territories 1996). High iron levels make it unsuitable for drinking, and there have been continuing controversies over salinisation of the resource due to forestry and mining on the Darling Range.

In arid areas, groundwater either from bores or from natural discharge points (mound springs) has been essential for the development of pastoralism. In most pastoral areas, there is now at least one water source within 10 km of most places (Figure 8.5). Unlike wild grazing animals, stock remain around the water points even during droughts, depleting the perennial vegetation. This means that plants which would be ungrazed by wild animals in poor seasons can be eaten out. Surveys at eight sites in the chenopod and acacia rangelands of southern and central Australia found little evidence of severe degradation but significant changes in species composition of plants and animals with distance from the watering points. The species which increased closer to water points were often exotics, whereas those which decreased closer to water points were mainly natives. This, and some signs of tree dieback close to the water points, indicate long-term degradation (Bennett 1997). A study between 1984 and 1992 in southern Northern Territory showed that a trend to increasing numbers of plant species with increasing distance from water was found on only some occasions and at some locations, not throughout the study period and areas. Nevertheless, vegetation was less diverse and more affected by variation in rainfall, season and grazing in heavily grazed sites, than in sites which had had

Figure 8.5
A road train pulled into Davenport Downs cattle station in the Channel country of southern Queensland. (Source: photo by R. W. Young.)

lighter grazing. When grazing was reduced, the degraded sites stabilised but did not recover, apparently because of soil and landscape changes which are not reversible within years to decades (Friedel 1997).

The results were strongly dependent on the scale of measurement—'spatial variation in vegetation composition was more strongly controlled by environmental variation than by distance from watering points or cattle activity, at the scale at which measurements were made' (Friedel 1997, p. 148). At first, measurements were made using point counts of vegetation along a presumed grazing gradient away from the watering point. These did not show a relationship to grazing impacts. Then 1 m^2 quadrats were used, and gave better ability to detect changes. However, over the length of the measured gradient, local variations in habitat over-rode any changes that may have been due to grazing. After heavy rainfall, the pattern of grazing altered, as stock ranged more widely. This also affected the assessment of the impact of grazing. In short, both changes in the patterns of grazing over the study period and the variations such as soil type and erodibility along measured gradients must be understood or the apparent impacts of grazing may be misrepresented. As discussed in Chapter 2, we need to be careful about interpreting environmental data, and to avoid seeing in experimental results only those 'facts' we believe.

Conclusion

The use of water, like the use of land, raises complex questions not only about the physical nature of the resource and how this can be affected by usage, but also about the social, economic, political and cultural frameworks in which the resource is managed. In a thoughtful review of this topic, Day (1996, p. 40) comments:

> Frequently, 'environmental management' and environmental change in nature have nothing to do with one another. Society 'manages' water for particular goals such as water supply to cotton irrigation or for cities or for dilution of salt . . . Yet while many rivers are 'managed' to change appreciably due to river regulation below a dam, the environmental impacts may not be managed at all . . . While current water policy supports our current quality of life goals, it cannot be taken to underpin 'sustainability' for the next century, despite purporting to do so.

This consideration returns us to questions of perceptions about our natural landscape and of the wisdom of the frameworks within which we modify them. These questions will be taken up again in the concluding chapter.

Chapter 9

Nature Conservation

Introduction

THE MODIFICATIONS made to Australian landscapes since 1788 have been wide-ranging and often drastic. However, it is hard to see how the population of Australia could have grown to the extent it has and become so affluent without such changes, especially given the scientific understanding, social attitudes and technology that have prevailed throughout the two hundred years of white settlement:

> Australia's geographical position and geological evolution make this continent one of the least productive regions of the world. The climates are too arid and variable, the soils too old and infertile, and the adjacent seas too low in nutrients, for high production per unit of land or sea. Historically, European attitudes to Australian natural resources have been chronically over-optimistic. They have concentrated on the small population and apparently unlimited, cheap space, without assessing its real potential. Major shifts of ecosystems have taken place since European settlement. It would not be possible to return all environments to a pristine pre-European condition, even if this were considered desirable. Invasions by successful alien species, together with great changes to vegetational composition, have altered many Australian environments permanently.
>
> Hamblin 1991, p. 2

Recognition of these facts in recent years has had two significant and related consequences: first, a drive to conserve remaining pristine areas and, second, an emphasis on sustainability in the future development of resources.

As discussed in Chapter 1, the early national parks were seen more as urban recreation parks than as conservation areas. However, even in the late nineteenth century and into the early twentieth century, many people did appreciate the natural flora and fauna, and scenic values. This can be seen in the emergence of bushwalking clubs, and societies dedicated to wildlife preservation and the protection of 'primitive areas'. However, most conservation areas have been declared since the 1980s (Table 9.1). Apart from cultivated areas, most of Australia's lands have been maintained under government control. Most of the large grazing properties in the semi-arid and arid regions are leasehold; forests are largely State-owned; water rights are licensed by the States; and roads and railways are public property. As demand for better conservation has replaced an ethic of 'wise use' in the management of natural resources, changes in the allocation of public lands has been possible, and public ownership has allowed large areas to be set aside for conservation purposes (Frawley 1988).

Table 9.1
Expansion of terrestrial conservation areas in Australia since 1968 (area in thousands of ha)

	1968	*1978*	*1986*	*1988*	*1996*
Australian Capital Territory	5	10	112	112	123
New South Wales	862	2 073	3 439	3 812	4 274
Victoria	201	295	1 401	1 823	3 406
Queensland	941	2 212	3 492	3 664	6 645
South Australia	1 170	3 921	6 711	11 117	20 957
Tasmania	288	681	948	967	2 177
Western Australia	1 151	12 649	14 649	15 252	15 927
Northern Territory	4 843	5 223	3 779	4 000	2 954
External territories	n.d.	n.d.	3	3	n.d.
Commonwealth					3 284

n.d. = no data
In total, Australia had 59.7 million ha of terrestrial reserves and 38.9 million ha of marine reserves in 1997. Data from Frawley 1988, ABS 1992, Cresswell and Thomas 1997. Note that some apparent discrepancies may be due to transfer of land from State to Commonwealth control. The Commonwealth total includes 1.15 million ha of areas protected under the Antarctic Treaty. Data prior to 1996 refers to areas within states; data for 1996 refers to the controlling jurisdiction.

Even this great expansion has not ensured a fully representative system of reserves. Also, clearance of bushland on private property is largely uncontrolled, as only Victoria and South Australia have attempted to legislate against this (Kirkpatrick 1994). There are scientific problems that need investigation, such as the best methods and the appropriate rationale for designing a system of reserves. There are also semantic problems—what do we mean, for instance, when we use words such as 'naturalness' or 'wilderness'?

'Naturalness': Worthless, priceless or indefinable?

The most dramatic changes in attitude towards the Australian environment have been recent—since about 1970—and new laws have formalised these altered viewpoints: laws to control pollution, to ban or restrict the use of chemicals that affect global climate or ocean waters, to protect native flora and fauna, to save endangered species, to set aside wilderness, to regulate the management of resources such as forests. It is fashionable to talk about converting from an 'anthropocentric' view, in which natural resources are the servant of people, to a 'biocentric' view, in which people are simply part of the natural environment and have no special claim to supremacy. In Western civilisation, nature was not seen as separate from the God who created it, until the eighteenth century. Wilderness was seen as threatening, and as a symbol of sinful disorder, as indeed it has been seen in most civilisations. To cultivate land was to bring order and productivity, and to make the land more habitable for people. It was believed that this was pleasing to God, who was the ruler of both land and people (Glacken 1976). Now, however, the hierarchy that places God above people, who in turn are placed above nature, seems to many to be an anachronism, and nature is seen to be at least equally as important as people:

> There appeared about 1700 a positive revaluation of Nature in the form of a new kind of natural garden, and in the new appreciation of mountains and wilderness . . . From being a surrogate for God, Nature came to replace God in actual expression . . . The rise of Nature as a central concept for debate seems to be part of the secularisation of society proceeding rapidly from the Age of Enlightenment, through the increasingly impressive discoveries of science, to the present day when many thoughtful people find their essential meanings in Nature rather than in God. This is not without its geographical repercussions; in the form of land use, these emerge as national parks and nature reserves, and in economic life in the form of environmental impact statements and restrictions on development. Tourism has grown to embrace the visiting not only of the cultural shrines, but also of natural shrines.
>
> Jeans 1983, pp. 172 and 171

Yet words like 'wilderness' and 'nature' are not clear descriptors; they are value-laden words, and the meanings that they carry are not constant. During the last three hundred years Western societies have regarded nature in a variety of ways (Jeans 1983):

- as a commodity, or as a lifeless thing, either because Judaeo-Christianity makes it subordinate to humankind, or because it is reducible by science to measured criteria and observable facts
- as something separate and noble—an escape from the decay and dirt of civilisation
- as a philosophical principle that pervades society, provides a framework for action, and gives us terms such as 'natural rights' and 'natural justice'.

In the 1990s we could add a further category:

- as religion in itself, personified as Gaia (the modern version of Mother Nature) and providing a new framework for environmental improvement:

> Gaia [is] a compellingly simple idea that the Earth . . . is a biological superorganism capable of regulating its conditions to optimise the chance of survival . . . [and] an alternative to the notion that humans are central to the universe . . . For some of course, Gaia has become a religion. But it need not be frightening to scientists. The word, religion, means to bind again . . . As far as the human species is concerned, that is the priority—to bind ourselves again to the workings of life on earth. Whether we come to it by religion, by politics or by science doesn't really matter.
>
> Porritt 1991, p. 41

Whichever view we take, there is no doubt that our perception influences our management of natural resources. The *Australian Heritage Commission Act 1975* (Cwlth) allows 'natural' as well as 'cultural' places of 'aesthetic, historic, scientific or social significance, or other special value' to be listed on the Register of the National Estate. The criteria for listing are explicitly value-based; significance cannot be measured quantitively in the same way as, for example, concentrations of pollutants in water. If a landscape is assessed using a fixed rating scale or system—for example, by assigning higher rating to running water than to still water, or to water views than to land views, or to native vegetation than to cleared pasture—this procedure is still 'simply quantifying one's prejudices, not establishing significance' (O'Brien and Purdie 1990). No completely objective and independent assessment is possible, because as Taylor (1990) emphasises, 'naturalness' is a culturally constructed concept.

To the first Europeans arriving in Australia, the landscape seemed natural. In contrast to the European landscape, where human impacts were so obvious, there was minimal apparent evidence of land management, or of permanent land occupancy. While we now recognise the long history (over 40 000 years), and significant impacts, of Aboriginal occupancy, impacts over the last two hundred years have been of a different character:

> There are points, in both space and time, where humanity becomes a special agent of landscape change, but this special agent is still a natural one . . . Humanity's role in landscape change cannot be modelled either as a simple dichotomy (i.e., natural landscape/ecosystem versus cultural landscape/ecosystem), or as a simple progression . . . (e.g., natural, sub-natural, semi-natural, near-natural, cultural landscape/ecosystem) . . . The effects of human intrusion and influences do differ in degree through space and time, but they also differ in kind. Thus human-generated landscapes are marked by discontinuities where humanity becomes a special agent within nature due to the qualitative differences between the character of its impact on the landscape and the character of other natural impacts.
>
> Taylor 1990, p. 413

Australia has had two such discontinuities: one with the arrival of Aboriginal people, and the other with the arrival of Europeans. It is the latter discontinuity that has caused the most pervasive, significant and rapid changes to Australian landscapes. Yet some parts of Australia seem only slightly affected in comparison to others, and it is these areas that we are now seeking to conserve and to protect from human-generated change. In determining 'naturalness' for conservation purposes, the most common indicators are still the lack of apparent modification by human impact and the integrity of native vegetation cover. Thus many of the conserved areas are in places that were not considered useful for other purposes, particularly agriculture (Figure 9.1). Like the national parks declared in the USA, most of our reserves are 'worthless lands'. The boundaries of many national parks skirt irregularly around known mineral deposits and arable areas. The parks are given some token value by their use for tourism, but their existence 'has been as much dependent on the absence of material wealth as it has been on the weight of aesthetic and ecological arguments' (Hall 1988, p. 453).

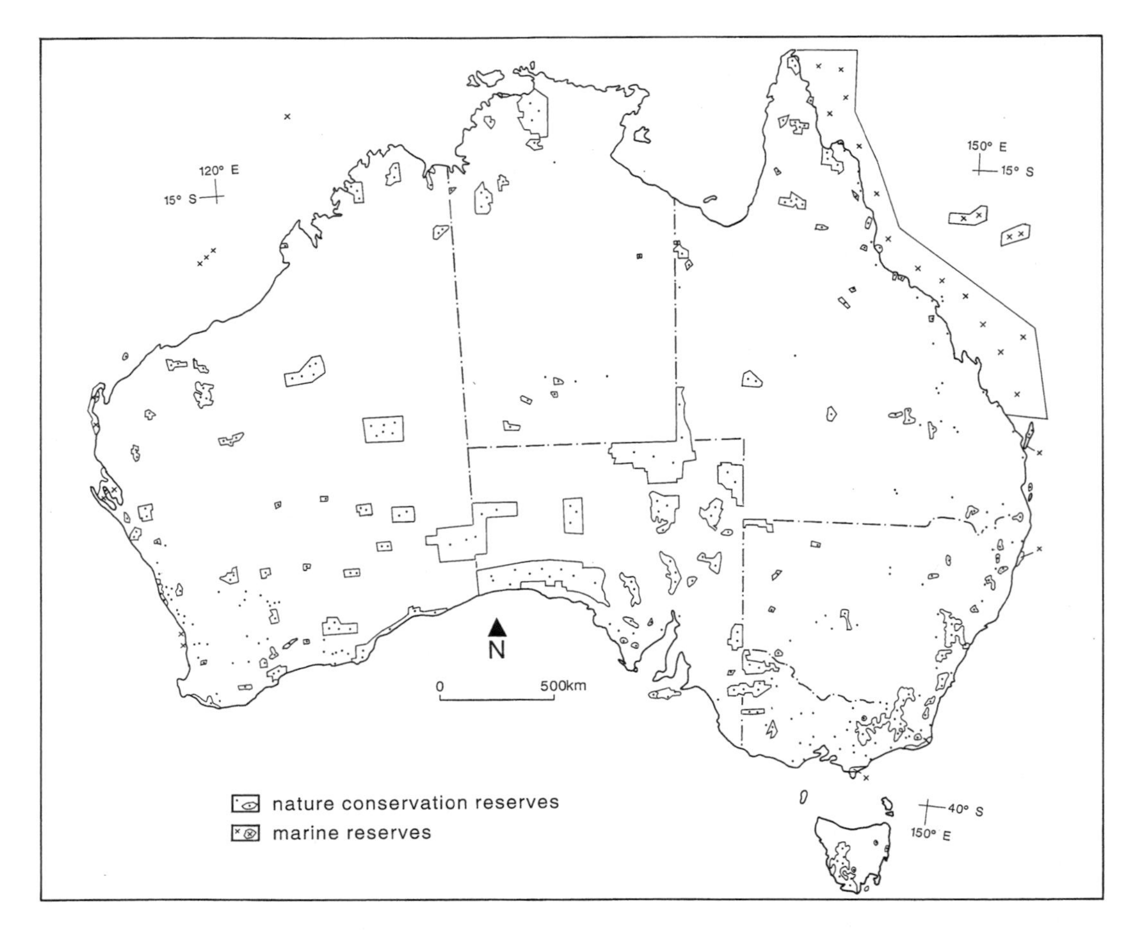

Figure 9.1 Nature conservation areas in Australia (Source: AUSLIG 1993.)

Wilderness and the impacts of tourism

> One of the most fascinating areas in the study of the relationship between humans and their environment is the changing concepts of 'wilderness'. Today the word is much used by the tourist industry in Australia and is given widely diverse and confused meanings which bear little relationship to the historical meanings, which all have in common a sharp edge of contact with nature, either terrifying at one extreme or benign and spiritually uplifting at the other.
>
> Frawley 1988, p. 402

The 'sharp edge of contact' mentioned by Frawley is a natural and perceivable boundary. Today, because they are no longer protected by the harshness of their environments or the community's fear of the dangers within them, wild areas are protected by legislation, and the boundaries around them are politically derived and sometimes seem arbitrarily drawn. Wilderness can be defined, using terms specified by the International Union for the Conservation of Nature

(IUCN), as an extensive area set aside primarily for ecosystem protection, but permitting minimal-impact recreation (Mercer 1993). The first surveys of wilderness quality in Australia were done in the late 1970s, but it was only in 1987 that the Australian Heritage Commission initiated a national inventory. The procedure used in the inventory recognised two essential attributes of wilderness: remoteness and naturalness. These are indicated by:

- remoteness from settlement (that is, from cleared land, or from permanently occupied sites, such as grazing property homesteads in arid lands)
- remoteness from access (that is, from constructed vehicle access routes, including aircraft landing strips, as well as roads and railways)
- aesthetic naturalness (that is, absence of permanent structures of modern technological society, including structures remote from settlement, such as powerlines, mining exploration sites, water supply bores and ruined buildings)
- biophysical naturalness (that is, absence of exotic plants and animals, or Australian plants and animals not endemic to the area, and absence of impacts due to other factors, such as water pollution derived from outside the area).

Remoteness was measured by estimating travel times from access routes or settlement, using times for four-wheel-driving in areas like the arid centre of the continent, and for bushwalking in areas like the rugged terrain of south-west Tasmania. Biophysical and aesthetic naturalness can be surveyed, but with all four criteria it was necessary to set arbitrary points of classification along two continuous spectra: from remote to settled, and from natural to human-dominated. Thus, even in this computer-based and quantitative assessment, the basis is value-laden and the quantification, to some extent, arbitrary. The value of the technique lies in its ability to provide a flexible approach to assessing the relative merits of competing claims on areas of varying wilderness quality, a process that is inherently political. For example, it does indicate that the areas of 'high' wilderness quality in Tasmania are largely located within either World Heritage areas or protected areas (Lesslie 1991) (Figure 9.2).

In 1990 the Land Conservation Council in Victoria—which had been set up in 1970 to oversee the use of public lands under the control of various authorities in the State—released a report nominating 23 areas of high wilderness quality for additional protection. After public discussion, most of the areas were protected under new wilderness legislation, but this did not greatly alter the conservation status of most of Victoria. In Australia's most densely settled State, less than 4% of the land was designated as wilderness; 90% of this was within existing national parks, and most of it was in the mallee of the north-western part of the State (Mercer 1993). Opposition to the declaration of wilderness came from an alliance of competing users: forestry operators, farmers, mining companies, cattlemen, four-wheel-drive clubs and game hunters. When New South Wales introduced similar legislation in 1994, there was strong opposition from horse-riders wanting to maintain access to the alpine areas. Horse-riding in alpine areas is likely to continue, and management of trampling damage and weed infestation is needed. Trampling causes immediate and persistent damage, especially in fens and bolster heath, but vegetation may regrow

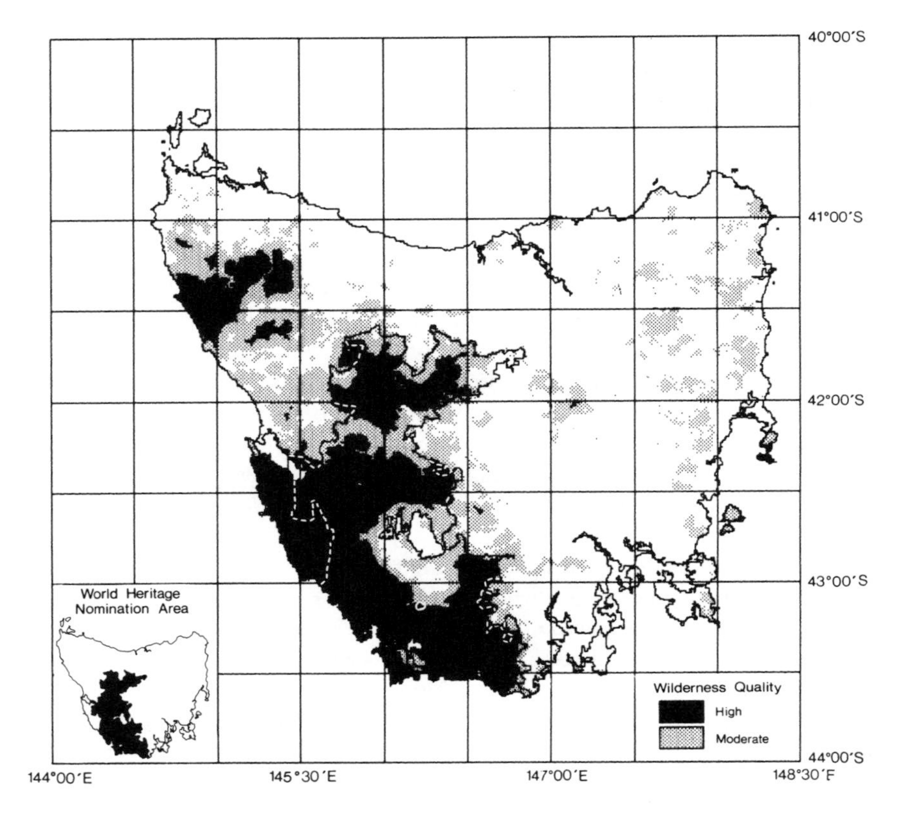

Figure 9.2 Wilderness quality in Tasmania. (Source: Lesslie 1991.)

if trampling does not persist. Weeds, particularly grasses, germinate in manured patches. Paradoxically, the exclusion of sheep from the World Heritage Area of the Central Plateau of Tasmania removes the grazers that most effectively remove these germinated weeds. However, the weeds are not strongly invasive and are likely to be grazed by native animals also (Whinam et al. 1994). Alpine areas, both in Tasmania and across the border between Victoria and New South Wales, continue to be a zone of tension in wilderness management, largely because of the competing requirements of recreation and tourism, and nature conservation (Box 15).

Demand for tourism and recreation in areas reserved for nature conservation is increasing, and inevitably has significant effects on the management of those areas. R. Beeton has commented on the shortcomings of Australia's national parks in this regard:

> In Australia, since 1965, the acquisition of protected lands has been driven by the biological representativeness principle . . . Since 1980, the wilderness criterion has been added progressively . . . Since the mid 1970s, historic and cultural areas have had a staggered start . . . In overall terms, the resultant system is a 'glad-bag' of coincidental national parks and other reserves with their location determined by criteria other than tourist needs or access routes. To create a system of greater value, the holes need filling and use conflicts need resolving.
>
> Beeton 1989, p. 46

In this assessment Beeton is not simply suggesting that the bases for selecting protected areas should be expanded from biological representativeness, wilderness, and cultural/historical value to include newer fashions in conservation such as biodiversity, unique geological sites, or heritage landscapes. He is suggesting something far more radical: that tourism is so strong a force in regard to protected lands that its needs should be considered when establishing reserves. Since the Native Title Act was passed in 1993, the importance of lands under the control of indigenous people has come into focus. Currently, indigenous people own or manage about 15% of the continent, and in the semi-arid and arid rangelands, the economic value of tourism and mining now exceeds the value of pastoral activities. In central and northern Australia, indigenous people are closely involved in the management of protected areas such as Uluru and Kakadu. The Aboriginal Land Council and the Cattleman's Union have agreed on land use strategies for Cape York, which take both natural/cultural values and economic issues into account. Thackway and Brunckhorst (1998) suggest that cooperative agreements for management like these provide hope for better environmental protection not only in protected areas and wilderness but also in the vast rangelands which occupy about 75% of Australia.

There is no doubt that ecotourism is growing in economic importance and that Australia's wild areas are a major drawcard for overseas tourists. Tourism is accelerated when an area is protected, particularly under the prestigious World Heritage label. However, the desires of tourists may clash with the objectives held for protected lands, so it is hardly surprising that it is difficult to achieve minimal impact in highly valued landscapes.

Box 15
Alpine wilderness on the Victorian–New South Wales border

Mainland Australia has a very small alpine area, which has been used for a wide range of activities: mining, recreation, nature conservation, water catchment, hydroelectricity generation, grazing, tourism, fishing, transport corridors, forestry. It is the home country of an Australian legendary character—the Man from Snowy River—and of all the cultural heritage that surrounds that legend. It is bisected by a state border, but management conflicts continue on both sides of the border. In a fragile environment, it is difficult to reconcile competing interests. Demand for ski facilities means clearing of vegetation and possible landslides. More accommodation generates larger volumes of sewage, which is difficult to treat when cold weather slows bacterial decay and releases nutrients into water bodies with little capacity to assimilate them. However, the management practices on either side of the border differ significantly, and the conservationist Geoff Mosley wrote in 1989 that 'both sides of the border are National Parks but the Kosciusko side is wilderness and the equally wild Victorian side is non-wilderness'. To understand this difference, we need to look at the differing histories of management.

The Aborigines used the alpine areas for food gathering, particularly when the Bogong moths swarmed, but they did not use fire extensively at high altitudes. Europeans settled the plains below the mountains from the 1820s, and drove their stock up to graze in summer. As cattle and sheep ate and trampled over the vegetation, soil was exposed and then washed away. Bare ground and tracks caused faster runoff, which led to severe erosion, and plants could not re-establish growth on bare rock or infertile subsoil. There had been periods of erosion prior to this, because of major fires that swept the ranges, but grazing and more frequent deliberate burning quickly caused both erosion and a change in vegetation to occur. Tree cover declined, and grasslands and herbfields spread. Weeds (including introduced pasture grasses) invaded, and many attractive native plants became rare.

On the New South Wales side conservation efforts began early. In 1906, a 206 km^2 park stretching from the Kosciusko summit to the Murray River was set aside for recreation and game preservation, and later, it was recognised that preservation of natural flora was important also. However, there was little real protection until the 5000 km^2 Kosciusko State Park was declared in 1944. This happened because of the enthusiasm of the bushwalker Myles Dunphy and the first director of the Soil Conservation Service, E. S. Clayton; these men involved the Premier, Sir William McKell, whose support for the park was strong and decisive. Concern about erosion had begun after the Hume Dam was constructed in 1936, and heightened when the Snowy Hydroelectric Scheme began in 1948. Grazing was increasingly restricted, until it was banned altogether in 1969 after the National Parks and Wildlife Service was established and took over park management. The High Country was occupied initially by gold miners and pastoralists, and then opened up for the Hydro-electric Scheme, but today tourism based on the ski fields and nature conservation dominates its economy. On tourist maps of the High Country, the power stations and dams are marked but the advertisements and illustrations highlight accommodation and recreational facilities.

Since the Kosciusko area has been national park, native vegetation has re-established itself on many once-eroded areas. Walking trails have been constructed to minimise the area disturbed by hikers, and erosion control measures have been used. Ironically, some early erosion measures were detrimental because zinc leached from the galvanised wire and retarded plant growth. Now inert materials, such as webbing made from old tyres and filled with local rock, are used (Figure 9.3, p. 193).

In Victoria the move towards conservation did not gain impetus until the 1980s. Dunphy had proposed reserving a Primitive Area on both sides of the border in 1935, but the idea was not pursued in Victoria, even after the Kosciusko Park was declared. Burning and sheep grazing were banned on the Victorian high plains after 1955, but often continued illegally. Ski resorts developed with few controls being placed on them. When four national parks were declared in the 1980s, grazing was withdrawn from the parks, but allowed to continue in other alpine areas. It was not until the Victorian wilderness legislation was passed in 1992 that many of the alpine areas achieved the level of protection that the New South Wales section had been given much earlier.

Data from Dodson et al. 1993, Mercer 1991, Mosley 1989, Johnston and Good 1996

Figure 9.3
Formed track near Blue Lake, Snowy Mountains

One of Australia's most popular bushwalking challenges is the Overland Track near Cradle Mountain in north-west Tasmania. Walkers crossing the peat mires create deep wallows (Figure 9.4). On steeper slopes and in more cohesive soils, deep furrows are worn, providing channels that are eroded to bedrock. Trampling destroys some plants, and poor hygiene near camp sites has led to outbreaks of gastroenteritis among walkers. S. S. Calais and J. B. Kirkpatrick (1986, p. 13) conclude that 'there is no vegetation in the park that will survive on paths used by 2000 people or more per annum', yet a greater number of people than this were using the track by 1980.

Foot traffic can also cause significant damage in arid environments. The Bungle Bungle Range in the eastern Kimberley region of Western Australia became a major tourist attraction in the mid-1980s. It is a visually striking landscape of convex 'beehive' towers and narrow sinuous ridges rising abruptly out of the surrounding plain. Algal skins give the sandstone towers and ridges a strongly contrasting horizontal banding of black and orange (Plate Q). The sandstone has a high compressive strength, but is weak in tension or shear. It is made up of closely interlocking grains of quartz, but there is almost no cementing material (such as clay) binding the grains together (Young and Young 1992). Thus, it can stand in steep faces, but it is easily broken when it is struck. The algal skins and occasional clayey veneers that bind the surface of the rock are thin and easily broken. Since they could damage the rock, even walkers are confined to the creek beds and banned from climbing the towers and ridges. Vehicles are excluded (Young 1987, Yulsman 1994), but remarkably, a number of television advertisements were made in which four-wheel-drive vehicles appear to be travelling recklessly along the creek bed of very similar terrain.

Figure 9.4 Bushwalkers in peat mud on the Overland Track, Tasmania, in 1993. (Source: photo by R. Delbridge.)

These two examples highlight the need for land managers to control and restrict public activities in order to conserve the value of the protected lands. Usually this is achieved using management plans that allocate varying uses across a park by some zoning system. The concept of the Recreation Opportunity Spectrum (Figure 9.5) is the basis of many of these plans. This concept recognises that each physical situation offers a range of possible opportunities for recreation, ranging from wilderness to completely altered developments. As a site is more and more developed, then numbers of visitors increase, the site becomes less natural, and more restrictions are placed on visitors' movements. Thus, a prominent lookout may become a developed location, with sealed access roads, a limited space for viewing, fenced and paved paths, planted gardens, and many associated facilities. A semi-natural opportunity within the same park may be provided by moving from the lookout along a narrow bush track across a ridge and into a sheltered valley not visible from the developed site. There are two criticisms of the use of the Recreation Opportunity Spectrum as a basis for zoning (van Oosterzee 1984):

- It takes only recreational needs directly into account. Other objectives, such as nature conservation, are only managed indirectly, by decisions to create 'natural' opportunities.
- It is not possible to provide the full range of opportunities in every area of protected land. The concept is more appropriate for regional planning, but it is often applied in the planning of relatively small parks.

The Great Barrier Reef Marine Park is perhaps Australia's premier example of regional planning of a protected area. Certainly it is our largest; it covers 345 000 km^2. The park is divided into four sections, each of which has a zoning plan (Box 16). Unlike most land-based parks, it is a multiple-use protected area; but fortunately, while economic uses are allowed, protection in perpetuity is the overriding goal. Tourism is the most valuable economic activity. In 1987/88 tourists on island resorts and charter vessels spent about $200 million; recreational fishing provided another $100 million, and commercial fishing output was $100 million also (Woodley et al. 1990). The effects of tourism include:

- impacts from construction of infrastructure (for example, dredging for breakwaters around marina sites)
- discharge of wastes, especially sewage, which causes eutrophication
- increased turbidity, due to dumping of dredge spoil, clearing on land for roads or buildings connected with tourism
- aesthetic impacts, particularly from floating offshore structures that cannot be screened by natural features
- damage to reefs when structures such as pontoons break free during storms or when large numbers of recreational fishing boats anchor at favoured sites
- disturbance to marine life (for example, by selective removal of preferred species by divers and fishers, or by activities such as fish and shark feeding).

Figure 9.5
The Recreation Opportunity Spectrum. (Source: van Oosterzee 1984.)

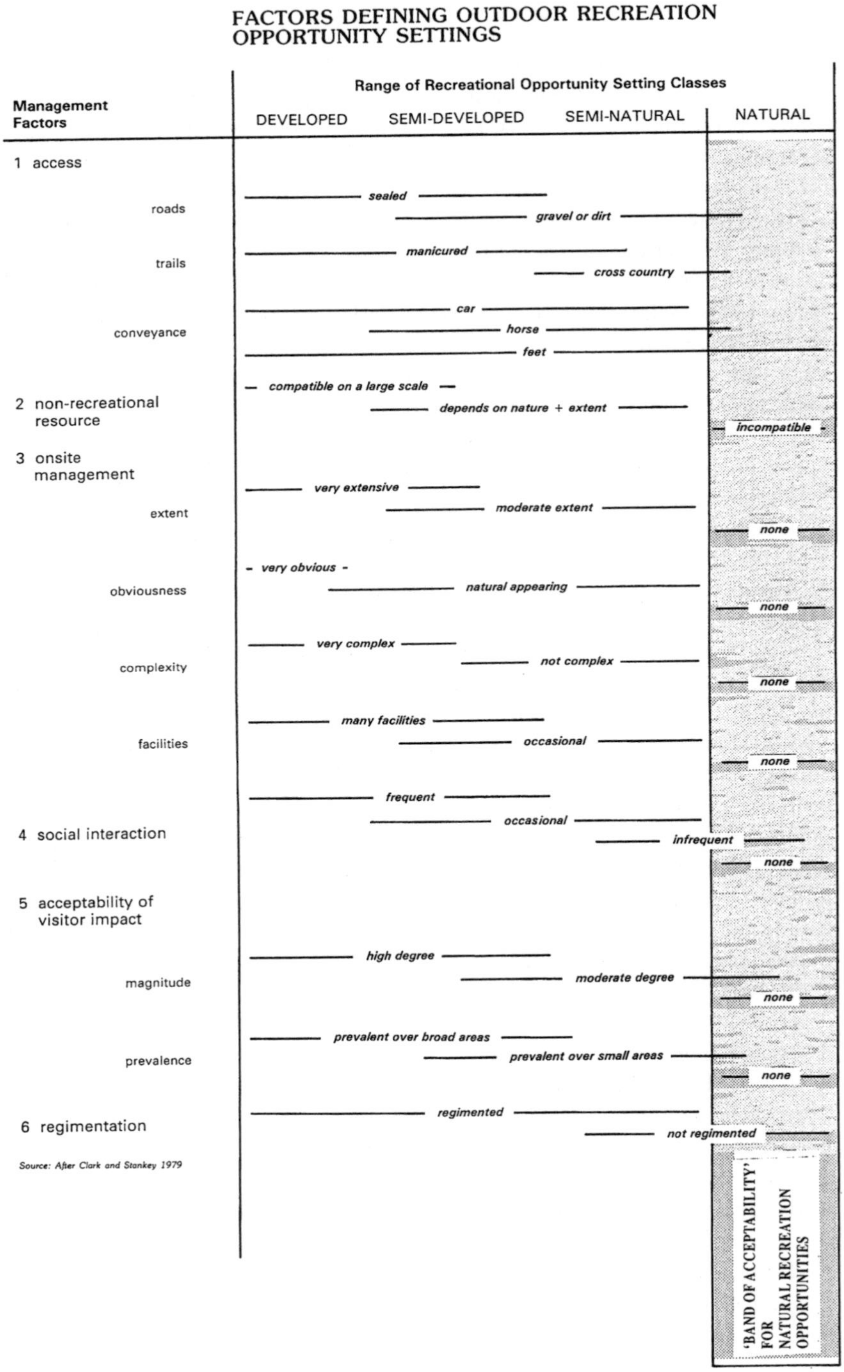

FACTORS DEFINING OUTDOOR RECREATION OPPORTUNITY SETTINGS

Management Factors

Range of Recreational Opportunity Setting Classes

DEVELOPED SEMI-DEVELOPED SEMI-NATURAL NATURAL

1 access

roads
sealed
gravel or dirt

trails
manicured
cross country

conveyance
car
horse
feet

2 non-recreational resource
compatible on a large scale
depends on nature + extent
incompatible

3 onsite management

extent
very extensive
moderate extent
none

obviousness
very obvious
natural appearing
none

complexity
very complex
not complex
none

facilities
many facilities
occasional
none

4 social interaction
frequent
occasional
infrequent
none

5 acceptability of visitor impact

magnitude
high degree
moderate degree
none

prevalence
prevalent over broad areas
prevalent over small areas
none

6 regimentation
regimented
not regimented

Source: After Clark and Stankey 1979

'BAND OF ACCEPTABILITY' FOR NATURAL RECREATION OPPORTUNITIES

Box 16
Zoning in the Great Barrier Reef Marine Park

Marine areas worldwide are poorly conserved, with less than 1% protected, although they occupy two and a half times the area of land on the globe. In terms of both the number of protected areas and the number of major biogeographic zones protected, Australia is a world leader, and zoning plans in the Great Barrier Reef Marine Park are in accordance with the integrated management regime favoured by the IUCN for marine reserves.

The Marine Park was established by Commonwealth legislation in 1975. It is administered jointly by the Queensland and Commonwealth governments, and the Act under which the reef ecosystem is conserved is a single piece of legislation that prevails over any conflicting legislation. There is also provision for regulations to control activities outside the park, such as nutrient discharges, that could affect the park adversely, but usually the Park Authority has preferred to negotiate change rather than use this provision. A zoning plan takes about two years to prepare. Draft plans are given to users of the zoned area, and there is a strong emphasis on public participation in the development of the final plan. Zoning policy has changed over time, from the early use of individual reefs as zoning units, to more recent attempts to use larger zoning units. This recent development recognises that many reef organisms produce larvae that move with the currents over considerable distances from their 'source' reefs to establish their adult forms on 'sink' reefs. For example, the 1992 rezoning in the Cairns region aimed to protect a connected set of 'source' and 'sink' reefs. Surveys indicate that a majority of people in all user groups perceive that zoning is successful in protecting the reef environment and is a wise use of public money.

The section of the reef under most pressure from economic uses is near Cairns (Figure 9.6). The zoning reflects this existing pressure, as well as established uses, onshore urbanisation and past impacts, all of which have altered the reef from its pristine state. Seventy-three per cent of the section is zoned for general use, 22% for habitat protection, and only 5% for more restricted purposes. In contrast, between Cooktown and Cape Melville there are substantial areas zoned for marine park, scientific research and preservation. In these zonings, activities such as fishing and tourism are much more restricted (Table 9.2).

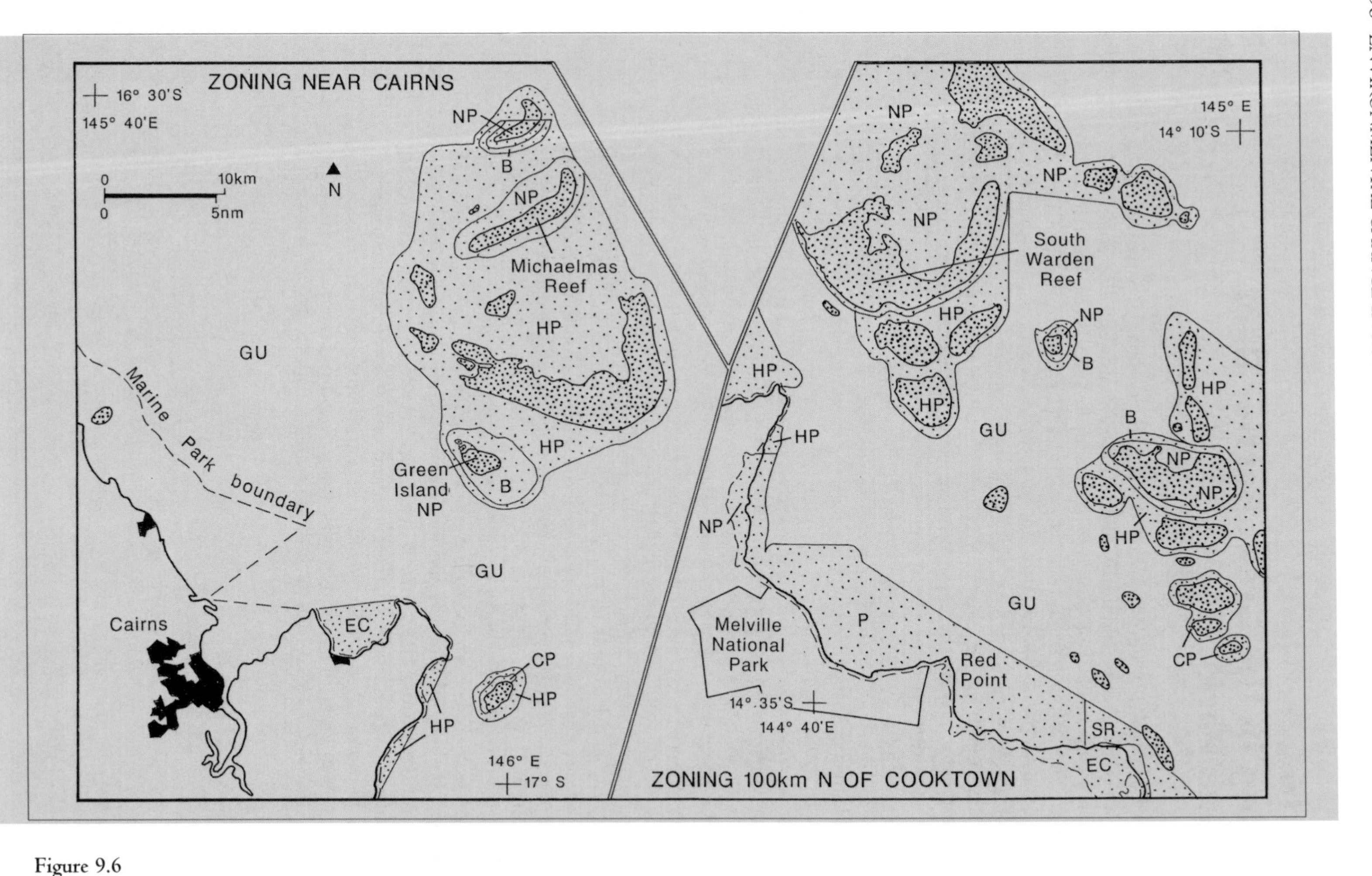

Figure 9.6
Zoning in the Great Barrier Reef Marine Park, near Cairns and Cooktown. The labels indicate the zonings, or the equivalent zonings, as listed in Table 9.2. The only exception is SR, which indicates Scientific Research zoning, under which only research and traditional fishing are allowed (and permits are needed for these uses). (Source: redrawn from Great Barrier Reef Marine Park Authority/Queensland National Parks and Wildlife Service 1992.)

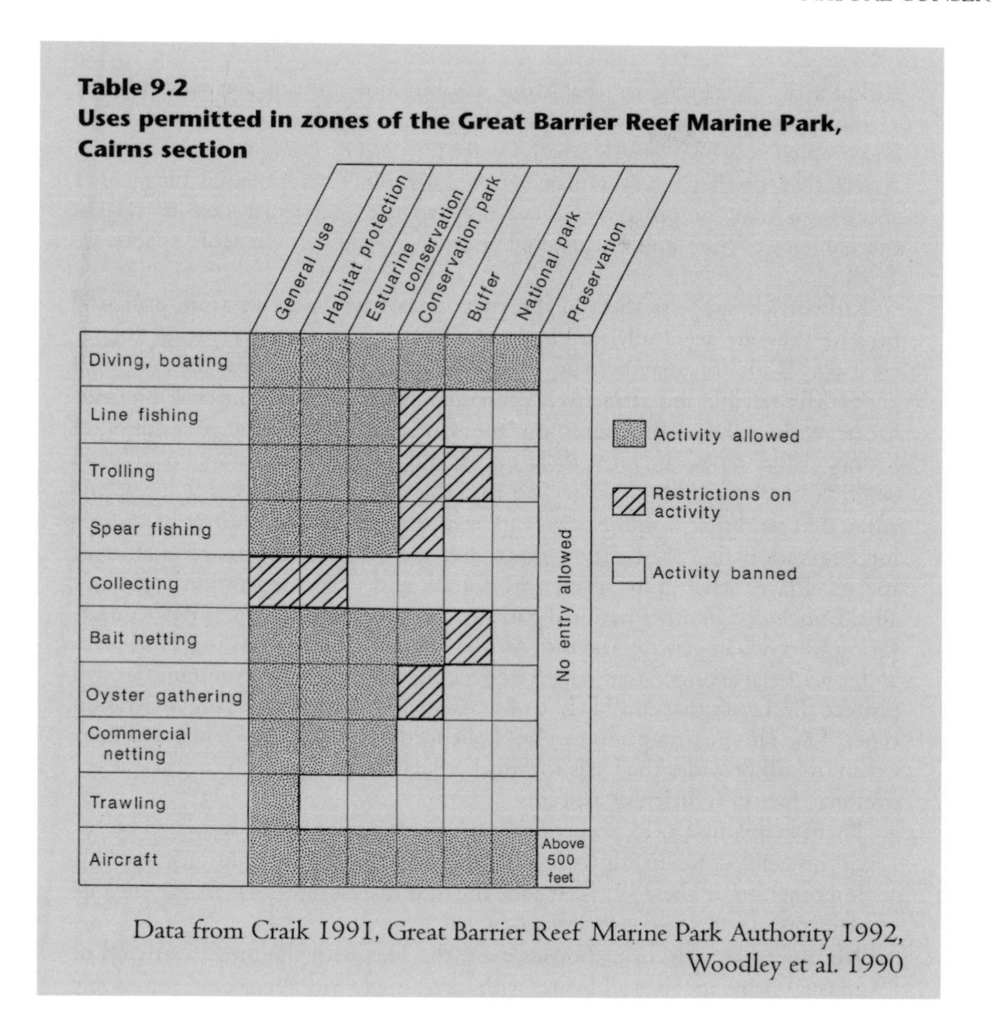

Table 9.2
Uses permitted in zones of the Great Barrier Reef Marine Park, Cairns section

	General use	Habitat protection	Estuarine conservation	Conservation park	Buffer	National park	Preservation
Diving, boating	Activity allowed	Activity allowed	Activity allowed	Activity allowed	Activity allowed	Activity allowed	No entry allowed
Line fishing	Activity allowed	Activity allowed	Activity allowed	Restrictions on activity	Activity banned	Activity banned	
Trolling	Activity allowed	Activity allowed	Activity allowed	Restrictions on activity	Restrictions on activity	Activity banned	
Spear fishing	Activity allowed	Activity allowed	Activity allowed	Restrictions on activity	Activity banned	Activity banned	
Collecting	Restrictions on activity	Restrictions on activity	Activity banned	Activity banned	Activity banned	Activity banned	
Bait netting	Activity allowed	Activity allowed	Activity allowed	Activity allowed	Restrictions on activity	Activity banned	
Oyster gathering	Activity allowed	Activity allowed	Activity allowed	Restrictions on activity	Activity banned	Activity banned	
Commercial netting	Activity allowed	Activity allowed	Activity allowed	Activity banned	Activity banned	Activity banned	
Trawling	Activity allowed	Activity banned	Activity banned	Activity banned	Activity banned	Activity banned	
Aircraft	Activity allowed	Activity allowed	Activity allowed	Activity allowed	Activity allowed	Activity allowed	Above 500 feet

Data from Craik 1991, Great Barrier Reef Marine Park Authority 1992, Woodley et al. 1990

How large do reserves need to be?

To protect ecologically valuable areas, and to conserve wilderness, most researchers argue that large reserves are necessary. The reasons are fourfold:

- Large areas are needed to maintain a minimum viable population, so that species are in high enough numbers, and have sufficient resources, to continue successful breeding.
- The theory of island biogeography is applied, picturing 'islands' of native vegetation in a 'sea' of developed land, and implying that, because the 'islands' are isolated, they need to be large to support a sustainable range of ecosystems.
- Protecting the genetic base—that is, the notion of evolutionary continuity—requires that biota can expand or contract their range if climatic or other environmental conditions alter.
- The larger the reserve, the smaller the ratio of its boundary length to its area and, consequently, the smaller the risk of invasion of exotic biota, or of degradation due to outside influences.

Kirkpatrick (1994) argues that these assumptions are not always valid. For example, a migratory bird whose home base is in a large reserve still needs other areas, which may be relatively small, for feeding and breeding when it migrates. A series of smaller reserves may better cover the environmental range of a species we want to protect, and we need to take particular care to reserve the habitats of rare and endangered species, even if the available spaces are small.

Kirkpatrick suggests that, in the past, areas have been set aside primarily because they are scenically striking and of no economic value except for recreation. While this may be true, it is remarkable that few Australian national parks offer reliable and attractively presented information on the striking landforms within them. Brochures on the sandstone-dominated parks around Sydney still describe the intricate lacework and pitting of solutional features as 'wind eroded caves', despite the fact that the floors of most caves are deeply covered in sand that would be easily moved by wind. The second set of criteria for reservation has been the rareness, and the current reservation status, of species. Places where rare species are found, and where vegetation types not already included in other national parks occur, have been sought as new conservation reserves. In general, the idea of 'the larger, the better' has been followed, and it has been assumed that conserving a range of vegetation communities will protect the fauna that are likely to be associated with the varying vegetation types. This latter assumption can be challenged (see Chapter 4), but to a large extent, it still provides the basis for much selection of reserves. However, selection may focus on different aspects:

- 'Richest area first' selection chooses the most diverse sites (measured by environmental domains, floristic communities, species distributions, or some combination of these). It then adds the next diverse sites that do not include elements already reserved, and so on.
- 'Minimum set' selection chooses, first, the sites with the most restricted or unusual elements, then adds sites with increasingly widespread and commonly occurring elements.
- 'At risk' selection gives preference to the most threatened elements.

These alternatives can be evaluated in different places to judge which offers the most cost-effective way of preserving biodiversity, and to plan a combination of reserves and management controls on other lands to achieve preservation of a diverse flora and fauna.

In many situations, however, adequate areas for reserves are simply not available. We are recognising that conservation and restoration of even small areas can be important in preserving biodiversity. Remnants of bushland within cities are the focus of bush regeneration programs. The 'long paddocks' of the travelling stock routes in country regions are refuges for species that have been cleared from adjacent cropping and pasture lands. Disturbed sites, such as mines, are being rehabilitated with endemic species, and some tree planting programs also use species that occur locally. Conservation of natural resources is not just a matter of setting aside a representative system of protected areas, but of managing all Australia's lands and seas; hence, the issue of sustainable development is clearly linked with the issue of conservation.

The concept of sustainability

The concept of sustainability is widely accepted today. It incorporates ideas of continued productivity, of some compatibility between development of resources and conservation of natural values, and of balance and rational decision-making. Yet it is a concept that has changed and developed over time (Table 9.3).

Table 9.3
Steps in the evolution of the concept of sustainable development, and the aims associated with each step

1930s Forestry in Australia
>>the concept of *sustainable use*
Aim:

- to manage forests to ensure a constant or increasing level of wood production in perpetuity

From 1964 The UN International Biological Program
Aim:

- to examine the biological basis of productivity and human welfare

1972 UN Stockholm Conference on the Human Environment
Aims:

- to safeguard and conserve natural resources, maintain renewable resources, share non-renewable resources
- to establish standards of environmental management, with international cooperation
- to regulate pollution, especially of the ocean
- to use science, technology and education to promote environmental management
- to link environmental concern and development, and give less developed countries incentives to promote rational management

1970s Man and the Biosphere Program
>>the concept of *ecodevelopment*, calling for radical changes to reduce social injustice as well as to protect the environment
Aims:

- to identify, monitor, and assess changes in the biosphere due to human activities, and the effects of these changes on humankind

1980 World Conservation Strategy
Aims:

- to maintain essential ecological processes and life-support systems
- to preserve genetic diversity
- to ensure sustainable utilisation of species and resources

In Australia, the National Conservation Strategy for Australia 1983 added:
- to maintain and enhance environmental quality

1987 ***Brundtland Report, Our Common Future***
>> the concept of *ecologically sustainable development (ESD)*
(This became government policy in Australia in 1990.)
Aims:
- to integrate economic and environmental goals in policies and activities
- to ensure that environmental assets are appropriately valued
- to provide for equity within and between generations
- to deal cautiously with risk and irreversibility
- to recognise the global dimension

1992 UNCED Earth Summit at Rio de Janiero
>> Principles embodied in the Agenda 21 action plans
Summary of the principles:
- Environmental protection is an integral part of development.
- The right to development must allow equity in the needs of present and future generations.
- The precautionary approach shall be widely applied.
- States shall cooperate to conserve, protect and restore the Earth's ecosystems, via arrangement for trade, international law, notification of disasters and transboundary pollution control.
- Effective legislation and participation by all concerned citizens is needed.
- States should reduce poverty, and unsustainable patterns of production and consumption, and should promote appropriate demography.
- Women, youth, and indigenous people should be given special consideration.
- Peace, development, and environmental protection are interdependent and indivisible.

1999 ***Environment Protection and Biodiversity Conservation Act*** **(C'w)**
- Replaces earlier Acts including the Environment Protection (Impact of Proposals) Act 1974 and the World heritage Properties Conservation Act 1983.
- Formally requires the principles of ESD to be considered in any project of national environmental significance.

Data from Kirkpatrick 1994, Moffatt 1992, Kelly 1992, Hawke 1990

The 1972 Stockholm Conference dealt with issues that were relevant to Australia: soil erosion, tropical ecosystem management, desertification, water supply, and human settlement. The statement issued after the conference recommended balancing conservation of natural resources and development of those resources. It also noted the special needs of less developed economies. These ideas evolved into the concept of ecodevelopment, which became a weapon in some less developed nations' fight against social injustice, and which has been seen by some governments as too radical in its approach. The World

Conservation Strategy of 1980, another United Nations initiative, was seen as an alternative to ecodevelopment and gained wider support. Including Australia, 40 countries adopted the Strategy and related it to their own conditions. In Australia this was the start of formal attempts to find consensus between development interests, such as mining companies, and conservationists. The Australian Government adopted the 1987 *Brundtland Report* and the concept of ecologically sustainable development (ESD) even more enthusiastically, although the Australian Conservation Foundation and Greenpeace argued that the government's response was weighted towards short-term economic interests. Nevertheless, the National Tree Planting Program, the National Soil Conservation Program, the declaration of World Heritage areas, the proclamation of the Decade of Landcare, and similar responses, can be seen as grounds for cautious optimism that the response is genuine and may lead to sustainable development (Moffatt 1992). The increasing role taken by the Commonwealth government is evident in these national programs and in the 1992 Intergovernmental Agreement on the Environment (IGAE). This agreement set up ground rules for interaction between all tiers of government on environmental issues and specifically provided for national criteria on matters such as environmental impact assessment.

At the Stockholm Conference in 1972, there were 113 nations represented, but only two heads of state were present. At the Rio Earth Summit, 20 years later, there were 178 nations and 107 heads of state or government. The Rio meeting was not so much a conference to discuss principles as it was an international negotiating forum. It set up conventions that would bind nations to take specific action—on climate change and biodiversity, for example. Non-binding action plans were brought together in Agenda 21, which dealt with consumption patterns and trade, a wide range of resource management issues, the role of groups such as women and indigenous peoples, and means of implementation such as technology transfer (Kelly 1992). The environment is an issue that features prominently on both worldwide and Australian political agendas, and 'sustainable development' is the catchcry for its management. However, it is obvious that, often, land management in Australia is not carried out on a sustainable basis:

> Most renewable resources on which primary industries depend are currently over-exploited . . . Many current practices in farming, forestry and fishing are not sustainable . . . Appropriate sustainable management practices will need to use a range of biological, economic and social indicators which are cast in relevant time-frames and value-sets.
>
> Hamblin 1991, p. 3

The question of a relevant time-frame is important. Many decisions that impact significantly on environmental management are made within a short time-frame: political decisions, crop planting and harvesting choices, and family financial decisions may be made almost annually. Bank loans, industrial technology changes, new product development, and regional planning frameworks may span one or more decades. Yet evidence of decline in the productivity of a

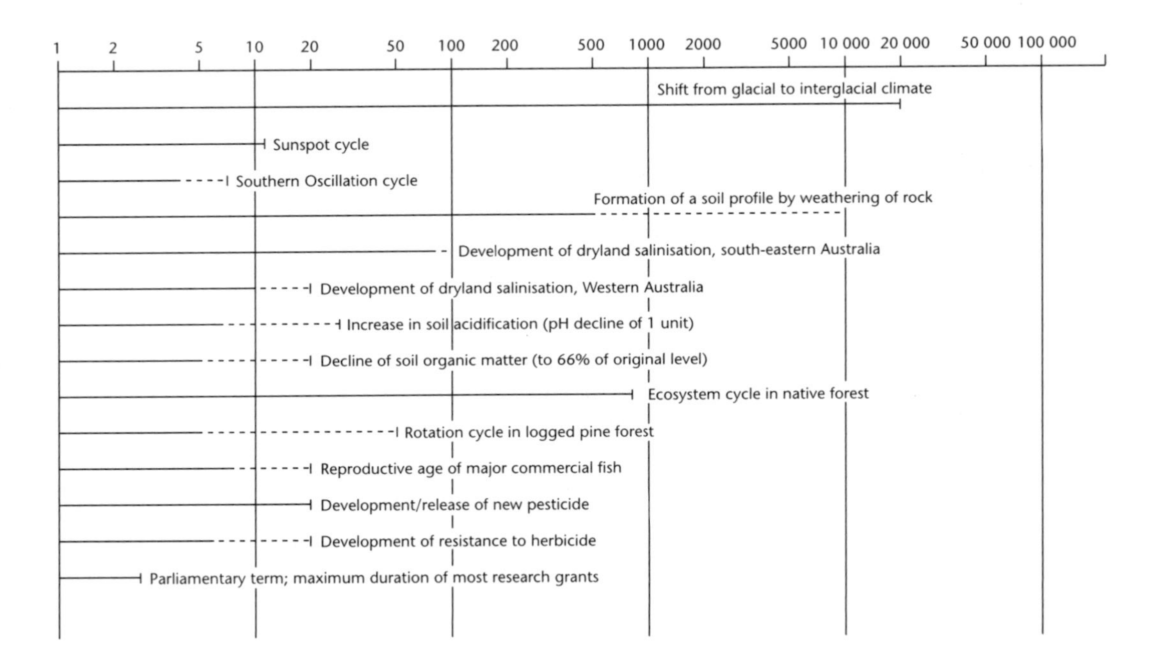

Figure 9.7 Time-line of environmental changes relevant to sustainable land use in Australia. (Source: data from Hamblin 1991.)

resource, or of fluctuations in natural forces such as climate, may become apparent over several decades, and renewing of eroded soil by weathering of exposed rock may take thousands of years (Figure 9.7). The advocates of ESD may talk of intergenerational equity and the need to take a long-term view, but in practice, this can be extremely difficult. Social and economic constraints are also important. For example, farmers with low net incomes are unlikely to allocate significant expenditure to land care. Farmers' attitudes (to risk taking, profit, or environmental management, for example) are important in explaining why people join Landcare groups when the idea is first introduced, but subsequent new memberships are explained better by situational variables (such as equity level, the educational qualifications of the farmer and off-farm income) (Black and Reeve 1993).

The Landcare movement is widely accepted by government bodies, as well as those involved in the movement, as one of the most hopeful signs that the Australian community is committed to sustainable development. Its growth has been rapid: from 150 groups in 1987, to 580 by 1990, and 1300 by 1992. In Western Australia, 48% of farmers are members; in Tasmania, where problems such as erosion or salinity are less obvious, only 10% of farmers have joined groups. Membership is expected to plateau at about 15% nationally. Landcare is supported by all political parties, by conservation and farming organisations, by major mining companies, and by all levels of government (Black and Reeve 1993). But can we speak of the 'Landcare movement' in the same way that we speak of the 'environmental movement'? Landcare has many small groups that are community-based: in farming areas, in schools, in local resident action associations. These are free from the direct control of government or Landcare administration, but they are indirectly and strongly influenced by these bodies because they are funded by competitive grants. These grants may be for projects such as erosion control or for administration to coordinate groups. Thus, the Landcare movement is more institutionalised and hierarchical than the environmental movement; the environmental movement is a social phenomenon, a large group of people with a strong sense of common identity, who may or may not belong to a formal environmental association such as the Australian Conservation Foundation or the Wilderness Society. S. Lockie (1992) argued that most farmers already share the ethic of land care—the concept of stewardship of the land so that it can be passed to the next generation in good condition. Yet 85% of farmers apparently see no advantage in belonging to a Landcare group; land management is poor not because farmers do not care, but because they have problems in perceiving some forms of land degradation, and in balancing economic survival and risk minimisation. The example of Landcare, and the differing perceptions that people have of the concept, illustrates the difficulties of translating obviously desirable and widely accepted concepts such as ESD into practical land management.

Precaution, politics or fashion? The greenhouse debate

The greenhouse debate began in the early 1980s with the argument that increased levels of carbon dioxide in the atmosphere, due to industrialisation and the burning of fossil fuels, were leading to warming of the earth and thence to climatic change and sea-level rise. Later, other gases were recognised as contributing to warming—notably methane, nitrous oxide, and halocarbons. At the Rio Summit in 1992 Australia argued for the *Convention on Climate Change*, which committed countries to reducing their emissions of greenhouse gases. This would have required Australia to stabilise its emissions at 1988 levels by 2000 and then reduce them. Industry and some of the government's advisers opposed this, because of serious flow-on consequences (Carruthers 1992, Jones 1992, Neck et al. 1992):

- Australia has 0.3% of the world's population and contributes only 1.5% of global anthropogenic carbon dioxide, but its major industries such as coal export would be severely affected.
- Price penalties on the use of coal for power generation would follow, yet there is no alternative renewable resource to replace coal and nuclear power generation is not acceptable in Australia.
- Enforcing higher prices for electricity could lead to adverse environmental consequences, such as pushing aluminium smelting offshore to places with less stringent environmental legislation, and the use of steel in vehicles, thus making them heavier and increasing fuel use.

At Kyoto in 1997, therefore, Australia argued successfully that it was one of a few countries which should be allowed to increase greenhouse emissions (relative to 1990 levels), whereas other industrialised nations should reduce their emissions by 5.2% by 2012. There was an outcry from conservationists but to date the government has maintained its position. Although 75% of carbon dioxide emissions are released by fossil fuel consumption, land clearing is the other major source. Carbon dioxide is released as vegetation is either burnt or left to rot. Not surprisingly, the government has targeted reduction of land clearing as an option for controlling greenhouse emissions. Unfortunately, as Lenzen (1998) points out, Australians are among the world's highest consumers per capita, due mainly to household purchases of goods and services. As incomes in Australia rise, so does consumption, and in conjunction with population growth, this outstrips reductions in greenhouse gas generation due to fuel mix changes or other technological improvements. Other strategies are required.

A major outcome of the Kyoto conference was acceptance of the concept of carbon offset trading. Reduction in greenhouse gases can be achieved either through direct cuts in emission, or by increasing carbon storage (mainly by tree planting). A Japanese motor vehicle company, Toyota, is establishing hardwood plantations in Western Australia, Victoria and South Australia, and gaining credit for the carbon stored in the trees which can be offset against the carbon dioxide released to the atmosphere by vehicle manufacture and use. BP Australia, which operates Western Australia's oil refinery, is paying that State's Department of Conservation and Land Management to plant trees for carbon credit; and in New South Wales, companies using coal for power generation are making similar arrangements with State Forests (Environment Institute of Australia 1998). Carbon credits like this may encourage individual farmers to plant more trees, by making plantings more financially attractive. This would help other environmental problems such as salinity.

More tree planting, less land clearing, use of renewable fuels, and improvements in fuel efficiency are all options identified by the government to reduce emissions, especially of carbon dioxide. Fermentation in landfills, and production of gas and coal are significant sources of methane. Small power stations have been set up to tap the methane from these places, using it to drive gas turbines. Methane from landfill typically produces less than 10 MW of power, but methane drainage from the Tower and Appin underground collieries in New South Wales is used to produce about 90 MW. In all these options, the stated aim is to foster sustainable use. Fact sheets produced by the Australian Greenhouse Office (1999) bear titles like those of publications from most government departments on any environmental issue—'Greenhouse sinks and sustainable land management', 'Efficient and sustainable energy use and supply'. The language indicates the impression which is sought, whether or not the substance of the material supports the indicated result.

Is all this activity and planning really necessary? While there is no dispute that carbon dioxide levels in the atmosphere have risen, the government acknowledges that observed climatic change could be attributed to natural climatic variability, rather than human-induced change. The popular view and the prevailing scientific view is that human-induced change is occurring, but it must be noted that not all evidence supports the warming result (Bryant 1987 and 1997, Daly 1989 and 1999):

- Some scientists proposed that the rise in carbon dioxide would lead to cooling and a new Little Ice Age. Events proposed as evidence for a warming globe—added snow and ice accumulation in Greenland, severe storms and floods in northern Europe in the late 1980s and early 1990s—could be seen equally as evidence for a cooling globe.
- Global mean temperatures declined about 0.3°C between the 1940s and 1980, having become about 0.45°C warmer between 1880 and 1940.

Surface temperatures have risen in the Northern Hemisphere but fallen in the Southern Hemisphere. Satellite measurements of temperature give different patterns to surface temperature measurements, suggesting that urban heat islands may be more significant than climatic change in causing the observed results.

- Worldwide, sea levels have risen in some places, dropped in approximately the same number of places, and been stable elsewhere.
- There were well-documented rises of temperature across the Northern Hemisphere beginning in the ninth century and peaking in the twelfth century, which allowed the Vikings to settle Greenland. Temperatures dropped by the fifteenth century, forcing abandonment of the settlements. Obviously, these significant changes predated the rise in carbon dioxide emissions. Estimates of temperature suggest that it rose by up to 2°C above the present level during the twelfth century (an increase comparable to greenhouse predictions), and fell to 1°C below the present level in the Little Ice Age of the late seventeenth century.
- Cynics suggested that the greenhouse effect and the problem of acid rain were being promoted as threats to the environment by advocates of nuclear power, anxious to discredit the use of fossil fuels for power generation.
- Some researchers believe that the recognised greenhouse gases have reached levels beyond which they can have no further effect in absorbing outgoing radiation and so will not cause further warming. Indeed, the major gas controlling absorption of radiation may be water vapour, over which we have no real control, rather than the gases so far identified.
- Climatic models are inadequate. They do not account adequately for dust emissions (both natural and human-induced), for changes in cloudiness or in cloud reflectivity (due, for example, to anthropogenic sulphate), for the effect of observed higher water vapour levels in the troposphere over the past 20 years, for natural changes in carbon dioxide emissions and ocean absorption due to ENSO-driven droughts.
- While we have reasonable estimates of carbon released globally by fossil fuel use and by clearing, we cannot account for all that is released. A significant percentage 'goes missing', perhaps into the biosphere, but potentially also into the other major storage, that is, the ocean.

Even if we accept that the modelled scenarios of a 2.5°C rise in temperature and significant shifts in climatic belts will occur, we need to be cautious about extrapolating the results. Almost any national parks brochure on endangered species will say that frogs are disappearing because of global climatic change (as well as habitat loss and other factors). The stated causes are either drier conditions due to greenhouse warming or increased UV radiation due to ozone depletion. Yet in alpine areas loss of suitable niches may be due more to drought than to greenhouse. Recent studies in Queensland rainforests are blaming a fungal disease, perhaps spread through the live pet trade (*Sydney Morning Herald* 20 March 1999)—but since this fungus may be favoured by warmer temperatures, some people may be tempted to return to the global warming cause. It is simply too easy to blame all environmental changes on a single cause. Studies of the vulnerability of nations to accelerated sea-level rise, initiated via the

Intergovernmental Panel on Climatic Change (IPCC), found that sea-level rise was often not as critical an issue as the consequences of poor planning or other factors, and that enhancement of existing problems needed to be considered (Harvey et al. 1999). Nevertheless, many climate researchers believe the evidence for greenhouse warming is convincing. And the consequences of warming are extensive: changes in agricultural productivity as climatic zones shift, flooding of coastal areas as the sea level rises, changes in tourism as ski fields shrink and some coastal areas gain milder weather, changes in cyclone frequency and storm damage. In determining its policy on greenhouse, the government invokes the precautionary principle of ESD. Since warming is supported by some evidence, and since greenhouse critics cannot prove that it will not occur, the wise course of action is to take steps to reduce as far as possible any adverse consequences.

The debate illustrates some of the difficulties in implementing the ideals of sustainable development. Firstly, as we have noted many times in earlier chapters, natural systems—particularly on a global scale, as in the case of climate—are complex, and it is difficult to be sure of cause–effect relationships. Secondly, actions to alleviate one problem—in this case, global warming—may trigger adverse impacts on other parts of the environment (for example, by encouraging the construction of large dams to provide alternative sources of power to fossil fuels). Thirdly, there are vested interests on both sides of the debate, and a truly objective approach to the issue is rare. Obviously action must be taken now, but only in the future will it be possible to look back and assess whether the decisions made about this, and other environmental controversies, were wise precautions against a serious threat, astute politics or blind following of environmental fashion.

Conclusion

The Australian environment seemed strange to the Europeans who came here in the late eighteenth century. They saw it as an empty and untouched country that needed to be altered to make it productive and habitable. We now acknowledge that it was not *terra nullius*: the Aboriginal people who had occupied the land for some tens of thousands of years had altered the landscape significantly (Dodson 1992), so that European settlement patterns were superimposed on a cultural, not a natural, landscape. Nevertheless, the landscape had then, and still has, larger areas that could be seen as 'wilderness' than the highly modified landscapes of Britain from which most of the settlers came. The new environment was not quickly understood or appreciated. For example, the push into physically unsuitable land across Goyder's Line in South Australia, the persistence of over-optimistic views about the potential for development in Western Australia and the continuing emergence of schemes to water the inland all illustrate how slowly we have come to recognise the limitations of our physical environment. Nevertheless, we should not imagine that environmental consciousness has arisen only in the last few decades. Recall that our earliest national park was set aside in 1879, that the difficulties of irrigation were predicted before the schemes began, and that writers in the nineteenth century were protesting against clearing, weed invasion and the spread of animal pests.

In attempting to understand the present environment and to make meaningful predictions of future environmental trends, we can liken ourselves to travellers on a typical outback Australian stream—a stream with several interlinking channels. We can look back along the bends of each of these channels and see that there have been significant changes, and variations in the rates of changes. For example, as we look back at the physical environment, we know that there have been important climatic changes that have moulded the present landscape: the dehydration of the continent since mid-Tertiary times allowing sclerophyll vegetation to replace rainforest over vast areas, and the shorter term variations in windiness, which partly explain variations in erosion over the past century. Or as we look back at human culture, we can see the two major discontinuities that occurred when Aboriginal, and then European, people arrived and modified the landscape. In addition to these major changes, there have been significant changes in attitudes to the landscape over the last two centuries, which have led to alterations in environmental management. And looking back along another channel—that of technology—important changes have resulted from technological innovations: the mechanisation of forestry, encouraging dieback and clearfelling; the introduction of legumes, leading to soil acidification; and the development of cars, allowing extended urban growth.

Even when we look back, we do not see everything. There are gaps in our knowledge, for example, about our poorly studied marine environments. Our understanding of physical processes may also change, as it did with the recognition that the sand supply on many beaches is restricted and not replenished by coastal rivers. The evidence we have may not give us conclusive answers to our questions; there are, for instance, still many uncertainties surrounding the ecological and long-term results of fuel-reduction burning in forests of different types.

The present environment and the way in which it is managed are contingent upon past physical, cultural and technological circumstances and changes. The better we understand these interconnected circumstances and changes, the better we will be able to make sensible decisions concerning future environmental management.

A Guide to Further Reading

This book has been organised mainly on the basis of the major categories of land use in Australia: agriculture, forestry, mining, coastal and marine activities, urbanisation, water resources and nature conservation. The environmental issues related to each of these activities have many aspects: the history of the land use in an area, the physical environment of the particular region, the political situation (both past and present), the socio-economic environment relevant to both an area and an industry. The following suggestions for further reading cover the most significant sources that were used in writing this book. Many of them include material that is relevant to more than one chapter. For a more comprehensive list of sources, see the References.

Jeans, D. N. (ed.) 1986, *Australia: A Geography*, vol. 1, *The Natural Environment*, Sydney University Press, Sydney.

Jeans, D. N. (ed.) 1987, *Australia: A Geography*, vol. 2, *Space and Society*, Sydney University Press, Sydney.

These two volumes provide excellent background material for all environmental issues. Chapters deal with aspects such as climate, soils, tropical environments, fisheries, population, transport, and recreation.

Heathcote, R. L. 1994, *Australia*, 2nd edn, Longman Scientific & Technical, Harlow, Essex.

Powell, J. M. 1988, *An Historical Geography of Modern Australia*, Cambridge University Press, Cambridge.

These books take a historical view of the development of the Australian landscape, and of changing attitudes. Both focus more on historical and socio-economic aspects than on physical processes.

Thomas, W. L. (ed.) 1956, *Man's Role in Changing the Face of the Earth*, University of Chicago Press, Chicago.

This is a pioneering work, published after an international symposium, and marking the start of geographers' strong involvement in environmental issues. Much of the material is still of interest.

Aplin, G. 1998, *Australians and Their Environment: An Introduction to Environmental Studies*, Oxford University Press, Melbourne.

Heathcote, R. L. (ed.) 1988, *The Australian Experience: Essays in Australian Land and Resource Management*, Longman Cheshire, Melbourne.

Rich, D. C. and Young, R. W. (eds) 1988, *Environment and Development in Australia*, Geographical Society of New South Wales, Sydney.

Russell, J. S. and Isbell, R. F. 1986, *Australian Soils: The Human Impact*, University of Queensland Press, St Lucia.

Saunders, D. A., Hopkins, A. J. M. and How, R. A. (eds) 1990, *Australian Ecosystems: 200 Years of Utilization, Degradation and Reconstruction*, Proceedings, Ecological Society of Australia, vol. 16, Surrey Beatty and Sons, Sydney.

These books deal with a wide range of environmental issues. Aplin's book gives a broad sweep and introduces the issues, but does not explore them at depth. The review publications produced for the 1988 Bicentennial (Heathcote, and Rich and Young) deal with both physical and social aspects, ranging from resource management to vegetation change to settlement patterns. Russell and Isbell, and Saunders et al., concentrate more on physical processes. However, all take a broad view. For example, Saunders et al. includes articles on attitudes to the environment, and chapters in Russell and Isbell discuss effects on soils as a result of urbanisation, mining and agriculture.

Hundloe, T. and Neumann, R. (eds) 1988, *Environmental Practice in Australia*, Environment Institute of Australia/Griffith University, Brisbane.

Mercer, D. 1995, *A Question of Balance: Natural Resources Conflict Issues in Australia*, 2nd edn, The Federation Press, Sydney.

Walker, K. J. (ed.) 1992, Australian Environmental Policy: Ten Case Studies, New South Wales University Press, Sydney.

These three books focus on environmental policy. They discuss specific environmental controversies and deal particularly with political, bureaucratic and social influences.

Barr, N. F. and Cary, J. W. 1992, *Greening a Brown Land: The Australian Search for Sustainable Land Use*, Macmillan Education Australia, Melbourne.

Frawley, K. J. and Semple, N. (eds) 1988, *Australia's Ever Changing Forests*, Special Publication No. 1, Department of Geography and Oceanography, Australian Defence Force Academy, Canberra.

Ivanovici, A., Tarte, D. and Olson, M. (eds) 1991, *Protection of Marine and Estuarine Areas: A Challenge for Australians*, Proceedings of the 4th Fenner Conference on Environment, Australian Committee for IUCN/Australian National Parks and Wildlife Service, Canberra.

Kirkpatrick, J. 1994, *A Continent Transformed: Human Impact on the Natural Vegetation of Australia*, Oxford University Press, Melbourne.

McTainsh, G. and Boughton, W. C. 1993, *Land Degradation Processes in Australia*, Longman Cheshire, Melbourne.

Smith, D. I. 1998, *Water in Australia: Resources and Management*, Oxford University Press, Melbourne.

The individual focuses of these books are clear from their titles: agriculture, forestry, marine and coastal areas, vegetation, land degradation and water. All provide detailed material on physical processes relevant to these issues, as well as discussions of important associated environmental controversies.

Ecologically Sustainable Development (ESD) Working Groups 1991a, *Final Report: Mining*, AGPS, Canberra.

—— 1991b *Final Report: Forest Use*, AGPS, Canberra.

—— 1991c *Final Report: Fisheries*, AGPS, Canberra.

Gifford, R. M. and Barson, M. M. (eds) 1992, *Australia's Renewable Resources: Sustainability and Global Change*, Bureau of Rural Resources Proceedings No. 14, Bureau of Rural Resources/CSIRO Division of Plant Industry, Canberra.

Resource Assessment Commission 1992, *Forest and Timber Inquiry Final Report: Overview*, AGPS, Canberra.

These are representative of the increasing range of major government reports on environmental issues. All provide excellent overviews of the current state of knowledge about the relevant issues, and of current policies concerning their management.

Department of Arts Sport and Territories 1996, *Australia: State of the Environment 1996*, CSIRO Publishing, Collingwood.

This state-of-the-environment report succeeds the ABS 1992 *Australia's Environment: Issues and Facts* and is representative of the range of such reports which are now being produced by the Commonwealth and all States, usually biennially. These are excellent sources of data and of interpretative material. Usually they are available on line as well as in hard copy.

The databases that are available on line or on CD-ROM are extremely valuable sources of information about environmental issues, particularly because they are constantly updated. The Heritage Australia Information Service (HERA) and the Environment Australia Online (ERIN on CD-ROM, or <www.erin.gov.au> on line) are good examples. A good starting place for web searching is the site for the relevant government. For Australian issues, try the Commonwealth Department of Environment and Heritage at <www.environment.gov.au>. Each state has a similar web page. For example, in Victoria, entry into the Department of Natural Resources and Environment (<www.nre.vic.gov.au>) will make available the EPA site. In New South Wales and Western Australia, access the environment department site directly (<www.epa.nsw.gov.au> and <www.environ.wa.gov.au>). Linked sites will allow access to other relevant web pages, such as forestry or agriculture department sites, or the current state-of-the-environment report. For current research material, the CSIRO web page (<www.csiro.gov.au>) is helpful.

References

ABARE (Australian Bureau of Agricultural and Resource Economics) 1999, <http://www.abare.gov.au/pubcat/AC/decoverviews.html> 29 April 1999.

ABS (Australian Bureau of Statistics) 1982–99, *Year Book Australia*, ABS Cat. no. 1301.0, ABS, Canberra.

—— 1992, *Australia's Environment: Issues and Facts*, Cat. no. 4140, ABS, Canberra.

—— 1994, *Population Growth and Distribution in Australia*, Cat. no. 2822.0, ABS, Canberra.

Agriculture, Fisheries and Forestry—Australia 1999, <http://www.affa.gov.au/agfor/forests/policy/index.html> 14 April 1999.

AGSO (Australian Geological Survey Organisation) 1998, 'The Life-sustaining Resources of the Piccadilly Valley (SA)', *AUS.GEO News* 48, October, p. 3.

Ahern, C., Weinand, M. and Murtha, G. 1992, 'Surface Soil pH in Non-arid Queensland', *Australian Journal of Soil and Water Conservation*, vol. 5, no. 2, pp. 38–43.

Allison, G. B., Cook, P. G., Barnett, S. R., Walker, G. R., Jolly, I. D. and Hughes, M. W. 1990, 'Land Clearance and River Salinisation in the Western Murray Basin, Australia, *Journal of Hydrology*, no. 119, pp. 1–20.

Anon. 1993, 'Iron Ore Resources', *Interface*, September, pp. 6–12.

Antarctic Division 1999, <http://www.antdiv.gov.au/environment/implement.htm> 27 April 1999.

Aplin, G. 1998, *Australians and Their Environment: An Introduction to Environmental Studies*, Oxford University Press, Melbourne.

Archer, M., Burnley, I., Dodson, J., Harding, R., Head, L. and Murphy, A. 1998, *From Plesiosaurs to People: 100 Million years of Australian Environmental History*, Australia: State of the Environment Technical Paper Series (Portrait of Australia), Department of the Environment, Canberra.

Australian Greenhouse Office 1999, revised 20 January, <http://www.greenhouse.gov.au/pubs/factsheets/index.html> 15 May 1999.

Australian Medical Association 1997, *Air Pollution and Health: The Facts*, Conference Papers, 10 November, Barton.

Australian Minerals and Energy Environment Foundation 1999, *Groundwork*, March, vol. 2, no. 3.

Australian Mining Industry Council 1992, *What Mining Means to Australians*, AMIC.

Australian National Parks and Wildlife Service 1990, *Ningaloo Marine Park (Commonwealth Waters) Plan of Management*, Australian National Parks and Wildlife Service, Canberra.

AUSLIG (Australian Surveying & Land Information Group) 1992, *The Ausmap Atlas of Australia*, Cambridge University Press, Melbourne.

—— 1993, *Australia Land Tenure*, Edition 1, 1:4 700 000, AUSLIG, Canberra.

Baghurst, P., Robertson, E., McMichael, A., Vimpani, G., Wigg, N. and Roberts, R. 1987, 'The Port Pirie Cohort Study: Lead Effects on Pregnancy Outcome and Early Childhood Development', *NeuroToxicology*, vol. 8, no. 3, pp. 395–402.

Bambrick, S. 1979, *Australian Minerals and Energy Policy*, Australian National University Press, Canberra.

Bardsley, K. L., Fraser, A. S. and Heathcote, R. L. 1983, 'The Second Ash Wednesday: 16 February 1983', *Australian Geographical Studies*, vol. 21, no. 1, pp. 129–41.

Barr, N. F. and Cary, J. W. 1992, *Greening a Brown Land: The Australian Search for Sustainable Land Use*, Macmillan Education Australia, Melbourne.

Bartle, J. R., Shea, S. R. and Kimber, P. C. 1981, 'Management of Tree Cover on Water Supply Catchments in Western Australia', in Old, K. M., Kile, G. A. and Ohmart, C. P. (eds), *Eucalypt Dieback in Forests and Woodlands*, CSIRO, Melbourne, pp. 212–26.

Bates G. M. 1987 *Environmental Law in Australia*, 2nd edn, Butterworths, Sydney.

Beale, B. and Fray, P. 1990, *The Vanishing Continent*, Hodder & Stoughton, Sydney.

Beattie, A., Auld, B., Greenslade, P., Harrington, G., Majer, J., Morton, S., Recher, H. and Westoby, M. 1992, 'Changes in Australian Terrestrial Biodiversity Since European Settlement', in Gifford, R. M. and Barson, M. M. (eds), *Australia's Renewable Resources: Sustainability and Global Change*, Bureau of Rural Resources Proceedings, no. 14, pp. 189–202.

Beckmann, G. and Coventry, R. 1987, 'Soil Erosion Losses: Squandered Withdrawals From a Diminishing Account', *Search*, vol. 18, no. 1, pp. 21–6.

Beeton, R. 1989, 'Tourism and Protected Landscapes', *Proceedings, National Parks and Tourism Seminar*, NSW National Parks and Wildlife Service/ Royal Australian Institute of Parks and Recreation, Sydney, pp. 33–49.

Bell, L. C. 1986, 'Mining', in Russell, J. S. and Isbell, R. F., *Australian Soils: The Human Impact*, University of Queensland Press/Australian Soil Science Society, St Lucia, ch. 18, pp. 444–66.

Bell, A. 1989, 'Meteorology and pollution in the Latrobe Valley', *Ecos*, vol. 43, Autumn, pp. 24–8.

Bennett, B. 1997, 'Water Points: Where Pastoralism and Biodiversity Meet', *Ecos*, vol. 92, winter, pp. 10–14.

Bergin, B. 1997, 'Development of a Methodology Allowing the Determination of Contamination Potential in Rural Areas', B.Env.Sc. (Hons) thesis, University of Wollongong.

Billen, G. and Lancelot, C. 1992, 'The Functioning of the Antarctic Marine Ecosystem: A Fragile Equilibrium', in Verhoeven, J., Sands, P. and Bruce, M. (eds), *The Antarctic Environment and International Law*, Graham & Trotman, London, ch. 7, pp. 39–41.

Birch, G., Shotter, N. and Steetsel, P. 1998, 'The Environmental Status of Hawkesbury River Sediments', *Australian Geographical Studies*, vol. 36, no. 1, pp. 37–57.

Bird, E. 1988, 'The Future of the Beaches', in Heathcote, R. L. (ed.), *The Australian Experience: Essays in Land Settlement and Resource Management*, Longman Cheshire, Melbourne, ch. 13, pp. 163–77.

Birtles, T. G. 1997, 'First Contact: Colonial European Preconceptions of Tropical Rainforest and Its People', *Journal of Historical Geography*, vol. 23, no. 4, pp. 393–417.

Bishop, P. and Li, S. 1997, 'Sub-basaltic Deep-lead Systems and Gold Exploration at Ballarat, Australia', *Australian Journal of Earth Sciences*, vol. 44, pp. 253–64. And discussion and reply, 1998, vol. 45, pp. 323–6.

Bishop, P. L. 1983, *Marine Pollution and Its Control*, McGraw Hill Book Company, New York.

Black, A. W. and Reeve, I. 1993, 'Participation in Landcare Groups: The Relative Importance of Attitudinal and Situational Factors, *Journal of Environmental Management*, no. 39, pp. 51–71.

Blainey, G. 1963, *The Rush That Never Ended*, Melbourne University Press, Melbourne.

—— 1993, 'Drawing Up a Balance Sheet of Our History', *Quadrant*, July–August, pp. 10–15.

Body, P. E., Inglis, G. R. and Mulcahy, D. E. 1988, *Lead Contamination in Port Pirie, South Australia*, Report no. 101, South Australian Department of Environment and Planning, Adelaide.

Bond, W. J. 1998, 'Effluent Irrigation: An Environmental Challenge for Soil Science', *Australian Journal of Soil Research*, vol. 36, pp. 543–55.

Bonyhady, T. 1985, *Images in Opposition*, Oxford University Press, Melbourne.

Boon, S. and Dodson, J. R. 1992, 'Environmental Response to Land Use at Lake Curlip, East Gippsland, Victoria', *Australian Geographical Studies*, vol. 30, no. 2, pp. 206–21.

Bowling, L. C. and Baker, P. D. 1996, 'Major Cyanobacterial Bloom in the Barwon–Darling River, Australia, in 1991, and Underlying Limnological Conditions', *Marine and Freshwater Research*, vol. 47, pp. 643–57.

Brennan, K. G., Noller, B. N., Le Gras, C., Morton, S. R. and Dostine, P. L. 1992, *Heavy Metals in Waterbirds from the Magela Creek Flood Plain, Alligator Rivers Region, Northern Territory, Australia*, Supervising Scientist for the Alligator Rivers Region Technical Memorandum 36, Australian Government Publishing Service, Canberra.

Bridgeman, H. A. 1989, 'Acid Rain Studies in Australia and New Zealand', *Archives of Environmental Contamination and Toxicology*, no. 18, pp. 137–46.

Bridgewater, P. and Ivanovici, A. 1991, 'Achieving a Representative System of Marine and Estuarine Protected Areas for Australia', in Ivanovici, A., Tarte, D. and Olson, M. (eds), *Protection of Marine and Estuarine Areas: A Challenge for Australians, Proceedings of 4th Fenner Conference on Environment*, Australian Committee for IUCN/Australian National Parks and Wildlife Service, Canberra, ch. 2, pp. 23–9.

Broderick, T. and Outhet, D. 1998, 'The Geomorphic Response of Receiving Streams for Inter-basin Transfer on the North Coast of NSW', abstracts, 8th Biennial Conference Australian and New Zealand Geomorphology Group, Goolwa, South Australia, 15–20 November.

Brodie, J., Arnould, C., Eldredge, L., Hammond, L., Holthus, P., Mowbray, D. and Tortell, P. 1990, *State of the Marine Environment in the South Pacific Region*, UNEP Regional Seas Reports and Studies No. 127, UNEP, Nairobi.

Brodie, J. 1997, 'Nutrients in the Great Barrier Reef Region', in Cosser, P. (ed.), *Nutrients in Marine and Estuarine Environments*, Australia: State of the Environment Technical Paper Series (Estuaries and the Sea), Department of Environment, Canberra, pp. 7–27.

Brooks, D. 1988, 'Environmental Management and Rehabilitation of Modern-Day Mine Sites', in Hundloe, T. and Neumann, R. (eds), *Environmental Practice in Australia*, Environment Institute of Australia/Griffith University, Brisbane, ch. 16, pp. 165–73.

Brown, C. 1991, *Pilgrim through This Barren Land*, Albatross Books, Sydney.

Brown, V. A., Smith, V. L. and Cox, A. 1991, 'Market Implications of US and OECD Proposals to Reduce Lead Exposure', *Agriculture and Resources Quarterly*, vol. 3, no. 2, pp. 194–202.

Bryant, E. 1987, 'CO_2-warming, Rising Sea-level and Retreating Coasts: Review and Critique', *Australian Geographer*, vol. 18, no. 2, pp. 101–13.

—— 1991, *Natural Hazards*, Cambridge University Press, Melbourne.

—— 1997, *Climate Process and Change*, Cambridge University Press, Cambridge/New York/Melbourne.

Buckney, R. T. and Morrison, D. A. 1992, 'Temporal Trends in Plant Species Composition on Mined Sand Dunes in Myall Lakes National Park, Australia', *Australian Journal of Ecology*, no. 17, pp. 241–54.

Burch, G. J. 1986, 'Land Clearing and Vegetation Disturbance', in Russell, J. S. and Isbell, R. F. (eds), *Australian Soils: the Human Impact*, University of Queensland Press/Australian Soil Science Society, St Lucia, ch. 7, pp. 159–84.

Bureau of Meteorology 1994, *Monthly Weather Reviews*, May.

Burgess, R. C., McTainsh, G. H. and Pitblado, J. R. 1989, 'An Index of Wind Erosion in Australia', *Australian Geographical Studies*, vol. 27, no. 1, pp. 98–110.

Bush, R. (ed.) 1993, *Proceedings, National Conference on Acid Sulphate Soils, Coolangatta*, 24–25 June, CSIRO/NSW Agriculture/Tweed Shire Council, Coolangatta.

Calais, S. S. and Kirkpatrick, J. B. 1986, 'Impact of Trampling on Natural Ecosystems in the Cradle Mountain: Lake St Clair National Park', *Australian Geographer*, vol. 17, no. 1, pp. 6–15.

Cameron, J. I. 1991, 'Land Degradation and the Wild Harvesting of Kangaroos', in Cameron, J. I. and Elix, J. (eds), *Recovering Ground: A Case Study Approach to Ecologically Sustainable Rural Land Management*, Australian Conservation Foundation, Melbourne, ch. 5, pp. 117–33.

Cameron, J. I. and Blick, R. 1991, 'Case Study 2: Pastoralism in the Queensland Mulga Lands', in Cameron, J. I. and Elix, J. (eds), *Recovering Ground: A*

Case Study Approach to Ecologically Sustainable Rural Land Management, Australian Conservation Foundation, Melbourne, ch. 4, pp. 75–115.

Cameron, J. I and Elix, J. (eds) 1991, *Recovering Ground: A Case Study Approach to Ecologically Sustainable Rural Land Management*, Australian Conservation Foundation, Melbourne.

Carnahan, J. A. 1986, 'Vegetation', in Jeans, D. N. (ed.), *Australia: A Geography*, vol. 1, *The Natural Environment*, Sydney University Press, Sydney, ch. 10, pp. 260–82.

Carne, R. J. 1993, 'Agroforestry Land Use: The Concept and Practice', *Australian Geographical Studies*, vol. 31, no. 1, pp. 79–90.

Carpenter, P., Butler, E., Higgins, H., Mackay, D. and Nichols, P. 1991, 'Chemistry of Trace Elements, Humic Substances and Sedimentary Organic Matter in Macquarie Harbour, Tasmania', *Australian Journal of Marine and Freshwater Research*, no. 42, pp. 625–54.

Carras, J. N. and Johnson, G. M. 1983. *The Urban Atmosphere: Sydney, a Case Study*, CSIRO, Melbourne.

Carron, L. T. 1985, *A History of Forestry in Australia*, Australian National University Press, Canberra.

Carruthers, D. 1992, 'The Earth Summit and the Risks it Poses for Australia's Economy', *The Mining Review*, vol. 16, no. 2, pp. 16–19.

Castle, R. J., Hewitt, J. W. and Suine, P. F. 1983, 'Mining Subsidence in Relation to Roads', pp. 377–97, in Knight, M. J., Minty, E. J. and Smith, R. B. (eds), *Collected Case Studies in Engineering Geology, Hydrogeology and Environmental Geology*, Special Publication No. 11, Geological Society Australia, Sydney.

Charman, P. E. V. and Murphy, B. W. 1991, *Soils: Their Properties and Management*, Sydney University Press/Oxford University Press, Melbourne.

Chartres, C. and Geeves, G. 1992, 'Soil Acidification in the Higher Rainfall Wheatbelt Zone of South-east Australia', *Australian Journal of Soil and Water Conservation*, vol. 5, no. 4, pp. 39–43.

Chenoweth, J. L. 1998, 'Conflict in Water Use in Victoria: Bolte's Divide', *Australian Geographical Studies*, vol. 36, no. 3, pp. 248–61.

Chittleborough, D. J. 1986, 'Loss of Land', in Russell, J. F. and Isbell, R. F. (eds), *Australian Soils: The Human Impact*, University of Queensland Press/ Australian Soil Science Society, St Lucia, ch. 20, pp. 491–511.

Churches, T. and Corbett, S. 1991, 'Asthma and Air Pollution in Sydney', *New South Wales Public Health Bulletin*, vol. 2, no. 8, pp. 72–3.

Clarke, C. J., Mauger, G. W., Bell, R. W. and Hobbs, R. J. 1998, 'Computer Modelling of the Effect of Revegetation Strategies on Salinity in the Western Wheatbelt of Western Australia: 2. The Interaction between Revegetation Strategies and Major Fault Zones', *Australian Journal of Soil Research*, vol. 36, pp. 131–42.

Cleland, K. 1992, *Residential Canal Development Proposed by Jimneva Properties P/L*, Report of the Commission of Inquiry for Environment and Planning (NSW), Commission of Inquiry for the Environment and Planning (NSW), Sydney.

Close, A. 1990, 'River Salinity', in Mackay, N. and Eastburn, D. (eds), *The Murray*, Murray Darling Basin Commission, Canberra, ch. 8, pp. 127–44.

Cohen, D., Martin, J., Bailey, G., Crisp, P., Bryant, E., Rothwell, R., Banks, J. and Hyde, R. 1994, 'A Twelve Month Survey of Lead in Fine Airborne Particles in the Major Population Areas of New South Wales', *Clean Air*, vol. 28, no. 2, pp. 79–88.

Commonwealth of Australia 1992, *National Forest Policy Statement: A New Focus for Australia's Forests*, AGPS, Canberra.

Conacher, A. J., Combes, P. L., Smith, P. A. and McLellan, R. C. 1983, 'Evaluation of Throughflow Interceptors for Controlling Secondary Soil and Water Salinity in Dryland Agricultural Areas of Southwestern Australia: 1. Questionnaire Surveys', *Applied Geography*, no. 3, pp. 29–44.

Cooke, B. D. (ed.) 1993, *Proceedings, Australian Rabbit Control Conference, Adelaide*, 2–3 April 1993, Anti-Rabbit Research Foundation of Australia, Adelaide.

Cosser, P. R. (ed.) 1997, *Nutrients in Marine and Estuarine Environments*, Australia: State of the Environment Technical Paper Series (Estuaries and the Sea), Department of Environment, Canberra.

Costello, L. and Dunn, K. 1994, 'Resident Action Groups in Sydney: People Power or Rat Bags?', *Australian Geographer*, vol. 25, no. 1, pp. 61–76.

Couper, A. 1990, *The Times Atlas and Encyclopaedia of the Seas*, Harper and Row, New York.

Crabb, P. 1988, 'Managing the Murray–Darling Basin', *Australian Geographer*, vol. 19, no. 1, pp. 64–88.

—— 1997, *Impacts of Anthropogenic Activities, Water Use and Consumption, on Water Resources and Flooding*, Australia: State of the Environment Technical Paper Series (Inland Waters), Department of Environment, Canberra.

Craik, W. 1991, 'The Great Barrier Reef Marine Park: A Model for Regional Management', in Ivanovici, A., Tarte, D. and Olson, M. (eds), *Protection of Marine and Estuarine Areas: A Challenge for Australians, Proceedings of 4th Fenner Conference on Environment, Canberra*, Australian Committee for IUCN/Australian National Parks and Wildlife Service, Sydney, ch. 13, pp. 91–7.

Cresswell, G. 1994, 'Nutrient Enrichment of the Sydney Continental Shelf', *Australian Journal of Marine and Freshwater Research*, vol. 45, pp. 677–91.

Cresswell, I. D. and Thomas, G. M. 1997, *Terrestrial and Marine Protected Areas in Australia*, Environment Australia, Department of Environment, Canberra (or <http://www.environment.gov.au/bg/nrs/protarea/paaust>).

CSIRO 1996, 'Don't Overfeed the Trees', *Ecos*, vol. 87, autumn, pp. 21–5.

—— 1998, salt supplement, *Ecos*, vol. 96, July–September, pp. 9–27.

CSIRO Marine Research 1999, 'Biomarkers: The Pollution Fingerprint', <http://www.marine.csiro.au/ResProj/CoastEnvmarPol/biomarkers.html> 16 April 1999.

Cunningham, D. M. 1988, 'A Rockfall Avalanche in a Sandstone Landscape, Nattai North, NSW', *Australian Geographer*, vol. 19, no. 2, pp. 221–9.

Daly, J. 1989, *The Greenhouse Trap*, Bantam Books, Sydney.

—— 1999, 'Still Waiting for Greenhouse', dated 13 April 1999, <http://www.vision.net.au/~daly/> 15 May 1999.

Dargavel, J. 1998, 'Politics, Policy and Process in the Forests', *Australian Journal of Environmental Management*, vol. 5, no. 1, pp. 25–30.

Davey, S. M. and Norton, T. W. 1990, 'State Forests in Australia and Their Role in Wildlife Conservation', in Saunders, D. A., Hopkins, A. J. M. and How, R. A. (eds), *Australian Ecosystems: 200 Years of Utilization, Degradation and Reconstruction*, Proceedings, Ecological Society of Australia, vol. 16, Surrey Beatty and Sons, Sydney, pp. 323–45.

Davidson, B. R. 1969, *Australia Wet or Dry?: The Physical and Economic Limits to the Expansion of Irrigation*, Melbourne University Press, Melbourne.

Davis, B. W. 1992, 'Antarctica as a Global Protected Area: Perceptions and Reality', *Australian Geographer*, vol. 23, no. 1, pp. 39–43.

Davis, G. 1993, 'Health Effects of Low-Frequency Electric and Magnetic Fields', *Environmental Science and Technology*, vol. 27, no. 1, pp. 42–51, and commentary, pp. 52–8.

Davis, W. J. 1993, 'Contamination of Coastal Waters Versus Open Ocean Surface Waters', *Marine Pollution Bulletin*, vol. 26, no. 3, pp. 128–34.

Dawson, P. and Weste, G. 1985, 'Changes in the Distribution of *Phytophthora cinnamomi* in Brisbane Ranges National Park Between 1970 and 1980–81', *Australian Journal of Botany*, no. 33, pp. 309–15.

Day, D. 1986, *Water and Coal*, CRES Monograph No. 14, Centre for Resource and Environmental Studies, Australian National University, Canberra.

—— 1996, 'Water as a Social Good', *Australian Journal of Environmental Management*, vol. 3, no. 1, pp. 26–41.

Department of Environment and Planning (NSW) 1980, *Resolution of Conflicts between Underground Extraction of Coal Resources and Dedication and Management of Areas as National Parks and Nature Reserves: Statement of Policy*, Department of Environment and Planning, Sydney.

Department of Environment Sport and Territories (Australia) 1996, *Australia: State of the Environment 1996*, CSIRO Publishing, Collingwood.

Department of Lands/Queensland Tourist and Travel Corporation 1992, *Sunmap Tourist Map: North Queensland*, edition 5, 1:750 000, Department of Lands, Brisbane.

Department of Resources and Energy (Commonwealth) 1983, *Water 2000: Consultants' Report No. 12*, AGPS, Canberra.

Derry, D. R. 1980, *A Concise World Atlas of Geology and Mineral Deposits*, Mining Journal Books, London.

Dick, A. 1998, 'Rivers of Unrest', *The Land* newspaper, Thursday 30 April, pp. 13–16.

Division of National Mapping 1986, *Atlas of Australian Resources, Third Series*, vol. 4, *Climate*, Division of National Mapping, Canberra.

Dixon, P. 1989, 'Dryland Salinity in a Subcatchment at Glenthompson, Victoria', *Australian Geographer*, vol. 20, no. 2, pp. 144–52.

Dixon, S. 1892, 'The Effects of Settlement and Pastoral Occupation in Australia upon the Indigenous Vegetation', *Transactions: Royal Society of South Australia*, vol. xv, no. 1, pp. 195–206.

Dodson, J. (ed.) 1992, *The Naive Lands*, Longman Cheshire, Melbourne.

Dodson, J., De Salis, T, Myers, C. A. and Sharp, A. J. 1994, 'A Thousand Years of Environmental Change and Human Impact in the Alpine Zone at Mt Kosciusko, New South Wales', *Australian Geographer*, vol. 25, no. 1, pp. 77–87.

Dovers, S. 1994, 'Recreational Fishing in Australia: Review and Policy Issues, *Australian Geographical Studies*, vol. 32, no. 1, pp. 69–114.

Dyson, P. R. 1983, 'Dryland Salting and Groundwater Discharge in the Victorian Uplands', *Proceedings, Royal Society of Victoria*, vol. 95, no. 3, pp. 113–16.

East, J. 1986, 'Geomorphological Assessment of Sites and Impoundments for the Long Term Containment of Uranium Mill Tailings in the Alligator Rivers Region', *Australian Geographer*, vol. 17, pp. 16–21.

Easton, C. 1989, 'The Trouble with the Tweed', *Fishing World*, March, pp. 58–9.

Eberbach, P. L. 1998, 'Salt-affected Soils: Their Cause, Management and Cost', in Pratley, J. and Robertson, A. (eds), *Agriculture and the Environmental Imperative*, CSIRO Publishing, Collingwood, pp. 70–97.

Eckersley, Y. 1998, 'The Problem with Blooming Algae', *Geo*, vol. 20, no. 1, January–February, pp. 36–40.

Ecologically Sustainable Development Working Groups 1991a, *Final Report: Fisheries*, AGPS, Canberra.

—— 1991b, *Final Report: Forest Use*, AGPS, Canberra.

—— 1991c, *Final Report: Mining*, AGPS, Canberra.

Edwards, A. 1949, 'Mineral Resources', in Wood, G. L. (ed.), *Australia: Its Resources and Development*, Macmillan, New York, ch. xiv, pp. 177–212.

Elix, J. and Cameron, J. 1991, 'Conclusions and Recommendations', in Cameron, J. I. and Elix, J. (eds), *Recovering Ground: A Case Study Approach to Ecologically Sustainable Rural Land Management*, Australian Conservation Foundation, Melbourne, ch. 10, pp. 205–18.

Ellis, R. C. and Lockett, E. 1991, 'The Management of High Altitude Eucalypt Forest in Tasmania', in McKinnell, F. H., Hopkins, E. R. and Fox, J. E. D. (eds), *Forest Management in Australia*, Surrey Beatty and Sons, Sydney, ch. 7, pp. 131–45.

Ellis, R. C. and Thomas, I. 1988, 'Pre-settlement and Post-settlement Vegetational Change and Probable Aboriginal Influence in a Highland Forested Area in Tasmania', in Frawley, K. J. and Semple, N. M. (eds), *Australia's Ever Changing Forests*, Special Publication No. 1, Department of Geography and Oceanography, Australian Defence Force Academy, Canberra, pp. 199–214.

Emmett, A. J. and Telfer, A. L. 1994, 'Influence of Karst Hydrology on Water Quality Management in Southeast South Australia', *Environmental Geology*, vol. 23, pp. 149–55.

Environment Institute of Australia 1998, *EIA Newsletter*, issue 15, August, pp. 4–5.

Environment Protection Authority (NSW) 1993, *NSW Government Lead Issues Paper*, EPA, Sydney.

—— 1995, *Sydney Deepwater Outfalls Final Report Series*, vol. 2, *Sewage Plume Behaviour*, EPA NSW, Sydney.

—— 1997, *NSW State of the Environment 1997*, EPA, Sydney, <http://www.epa.nsw.gov.au/soe/97/> 15 May 1999.

Environmental Defenders Office NSW 1992, *Environment and the Law*, CCH Australia, Sydney.

Erskine, W. D. and Warner, R. F. 1988, 'Geomorphic Effects of Alternating Drought and Flood Dominated Regimes on NSW Coastal Rivers', in Warner, R. F. (ed.), *Fluvial Geomorphology in Australia*, Academic Press, Sydney, pp. 223–44.

Erskine, W. D. and Warner, R. F. 1998, 'Further Assessment of Flood- and Drought-dominated Regimes in South-eastern Australia', *Australian Geographer*, vol. 29, no. 2, pp. 257–61.

Erskine, W. D., Warner, R. F., Tilleard, J. W. and Shanahan, K. F. 1995, 'Morphological Impacts and Implications of a Trial Release on the Wingecarribee River, New South Wales', *Australian Geographical Studies*, vol. 33, no. 1, pp. 44–59.

Evans, L. F., Weeks, I. A. and Eccleston, A. J. 1982, 'Source Areas and Impact Areas of Photochemical Smog in Melbourne', *Clean Air*, vol. 16, no. 3, pp. 45–54.

Evans, R., Brown, C. and Kellett, J. 1990, 'Geology and Groundwater', in Mackay, N. and Eastburn, D. (eds), *The Murray*, Murray–Darling Basin Commission, Canberra, ch. 5, pp. 77–93.

Ewan, C., Young, A., Bryant, E. and Calvert, D. 1993, *National Framework for Environmental and Health Impact Assessment*, report under the Commonwealth government's National Better Health Program, University of Wollongong, Wollongong.

Eyre, B. 1993, 'Nutrients in the Sediments of a Tropical North-Eastern Australian Estuary, Catchment and Near-Shore Coastal Zone', *Australian Journal of Marine and Freshwater Research*, no. 44, pp. 845–66.

Eyre, B. and France, L. 1997, 'Importance of Marine Inputs to the Sediment and Nutrient Load of Coastal-plain Estuaries: A Case Study of Pumicestone Passage, South-eastern Queensland, Australia', *Australian Journal of Marine and Freshwater Research*, no. 48, pp. 277–86.

Fisher, D. E. 1980, *Environmental Law in Australia*, University of Queensland Press, St Lucia.

Flannery, T. F. 1997, *The Future Eaters: An Ecological History of Australian Lands and People*, Reed New Holland, Sydney.

Foale, S. 1993, 'An Evaluation of the Potential of Gastropod Imposex as a Bioindicator of Tributyltin Pollution in Port Phillip Bay, Victoria', *Marine Pollution Bulletin*, vol. 26, no. 10, pp. 546–52.

Formby, J. 1987, 'Environmental Impact Assessment: Where Has it Gone Wrong?—The Tasmanian Woodchip Controversy, *Environmental Planning and Law Journal*, vol. 4, no. 3, pp. 191–203.

—— 1991, *Response to the Resource Assessment Commission Forest and Timber Inquiry*, Consultancy Report for the South-East Forests and Timber Inquiry.

Frawley, K. J. 1987, *Exploring Some Australian Images of Environment*, Working Paper 1987/1, Department of Geography, Australian Defence Force Academy, Canberra.

—— 1988, 'The History of Conservation and the National Park Concept in Australia: A State of Knowledge Review', in Frawley, K. J. and Semple, N. (eds), *Australia's Ever Changing Forests*, Special Publication No. 1, Department of Geography and Oceanography, Australian Defence Force Academy, Canberra, pp. 395–417.

Frawley, K. J. and Semple, N. (eds) 1988, *Australia's Ever Changing Forests*, Special Publication No. 1, Department of Geography and Oceanography, Australian Defence Force Academy, Canberra.

Friedel, M. H. 1997, 'Discontinuous Change in Arid Woodland and Grassland Vegetation along Gradients of Cattle Grazing in Central Australia', *Journal of Environmental Management*, vol. 37, pp. 145–64.

Gale, S. J., Haworth, R. J. and Pisanu, P. C. 1995, 'The ^{210}Pb Chronology of Late Holocene Deposition in an Eastern Australian Lake Basin', *Quaternary Science Reviews (Quaternary geochronology)*, vol. 14, pp. 395–408.

Galligan, B. and Lynch, G. 1992, 'Integrating Conservation and Development: Australia's Resource Assessment Commission and the Testing Case of Coronation Hill', *Environmental Planning and Law Journal*, no. 9, pp. 181–94.

Galloway, R. W. and Bahr, M. E. 1979, 'What is the Length of the Australian Coast?', *Australian Geographer*, no. 14, pp. 244–7.

Geary, P. 1992, 'Diffuse Pollution from Wastewater Disposal in Small Unsewered communities', *Australian Journal of Soil Water Conservation*, vol. 5, no. 1, pp. 28–33.

Gentilli, J. 1986, 'Climate', in Jeans, D. N. (ed.), *Australia: A Geography*, vol. 1, *The Natural Environment*, Sydney University Press, Sydney, ch. 1, pp. 14–48.

Gibbs, H. 1991, *Inquiry into Community Needs and High Voltage Transmission Line Development*, NSW Government Printer, Sydney.

Gifford, R. M. and Barson, M. M. (eds) 1992, *Australia's Renewable Resources: Sustainability and Global Change*, Bureau of Rural Resources Proceedings No. 14.

Glacken, C. 1976, *Traces on the Rhodian Shore*, University of California Press, Berkeley.

Gleeson, J. 1976, *Australian Painters*, Lansdowne Press, Sydney.

Government of Australia 1993, *Counter Memorial: Certain Phosphate Lands in Nauru (Nauru v Australia)*, to the International Court of Justice.

Graetz, D., Fisher, R. and Wilson, M. 1992, *Looking Back: The Changing Face of the Australian Continent, 1972–1992*, CSIRO Office of Space Science and Applications, Canberra.

Graetz, R. D., Wilson, M. A. and Campbell, S. K. 1995, *Land Cover Disturbance over the Australian Contintent: A Contemporary Assessment*, Biodiversity Series Paper No. 7, Department of Arts Sport and Territories, Canberra.

Graham, O. P. 1989, *Land Degradation Survey of NSW 1987–1988: Methodology*, Soil Conservation Service of NSW Technical Report No. 7, Soil Conservation Service of NSW, Sydney.

Grant, C. D. and Koch, J. M. 1997, 'Ecological Aspects of Soil Seed-banks in Relation to Bauxite Mining. II. Twelve Year Old Rehabilitated Sites', *Australian Journal of Ecology*, vol. 22, pp. 177–84.

Gray, D. 1991, 'Combating Oil Pollution of the Sea', *Search*, vol. 22, no. 8, pp. 285–6.

Great Barrier Reef Marine Park Authority/Queensland National Parks and Wildlife Service 1992, *Basis for Zoning: Great Barrier Reef Marine Park, Cairns Section, and the Cairns Marine Park*, GBRMP/QNPWS, Cairns.

Grenfell Price, A. 1972, *Island Continent*, Angus & Robertson, Sydney.

Hall, C. M. 1988, 'The ((Worthless Lands Hypothesis)) and Australia's National Parks and Reserves', in Frawley, K. J. and Semple, N. (eds), *Australia's Ever Changing Forests*, Special Publication No. 1, Department of Geography and Oceanography, Australian Defence Force Academy, Canberra, pp. 441–56.

Hamblin, A. 1991, *Sustainability: Physical and Biological Considerations for Australian Environments*, Working paper WP/19/89 (revised), Bureau of Rural Resources, Canberra.

Hamilton, A. G. 1892, 'On the Effect Which Settlement in Australia Has Produced Upon Indigenous Vegetation', *Journal and Proceedings, Royal Society of New South Wales*, no. xxvi, pp. 178–239.

Handmer, J. and Wilder, M. 1994, 'Australia and Environmental Management in Antarctica', *Australian Journal of Environmental Management*, vol. 1, no. 3, pp. 24–34.

Hanley, J. R. 1992, 'Current Status and Future Prospects of Mangrove Ecosystems in Northern Australia', in Moffatt, I. and Webb, A. (eds), *Conservation and Development Issues in Northern Australia*, North Australia Research Unit, Australian National University, Canberra, ch. 6, pp. 45–54.

Hannan, J. C. 1984, *Mine Rehabilitation: A Handbook for the Coal Mining Industry*, New South Wales Coal Mining Association, Sydney.

Harvey, N. 1996, 'The Significance of Coastal Processes for Management of the River Murray Estuary', *Australian Geographical Studies*, vol. 34, no. 1, pp. 45–57.

Harvey, N., Clouston, B. and Carvalho, P. 1999, 'Improving Coastal Vulnerability Assessment Methodologies for Integrated Coastal Zone Management: An Approach from South Australia', *Australian Geographical Studies*, vol. 37, no. 1, pp. 50–69.

Hawke, R. J. 1990, *Our Common Future*, AGPS, Canberra.

Heathcote, R. L. 1972, 'The Visions of Australia 1770–1970', in Rapoport, A. (ed.), *Australia as a Human Setting*, Angus & Robertson, Sydney, pp. 77–98.

—— (ed.) 1988, *The Australian Experience: Essays in Australian Land and Resource Management*, Longman Cheshire, Melbourne.

—— 1994, *Australia*, 2nd edn, Longman, London.

Heatwole, H. and Lowman, M. 1986, *Dieback: Death of an Australian Landscape*, Reed Books, Sydney.

Helyar, K. R., Cregan, P. D. and Godyn, D. L. 1990, 'Soil Acidity in New South Wales: Current pH Values and Estimates of Acidification Rates', *Australian Journal of Soil Research*, no. 28, pp. 523–37.

Henry, R. L., Hensley, M. J. and Bridgeman, H. A. 1989, *Asthma and Air Quality at Lake Munmorah and Nelson Bay*, Report to the Electricity Commission of New South Wales by the University of Newcastle.

Hills, B. 1989, *Blue Murder*, Macmillan, Melbourne.

Holmes, J. H. 1988, 'New Challenges Within Sparselands: The Australian Experience', in Heathcote, R. L. and Mabbutt, J. A. (eds), *Land Water and People*, Allen & Unwin, Sydney, pp. 75–101.

Holper, P. 1995–96, 'Acid Air? Tracking Acidification', *Ecos*, vol. 86, summer, pp. 30–4.

Holze, M. 1892, 'Introduced Plants in the Northern Territory', *Transactions, Royal Society of South Australia*, vol. xv, no. i, pp. 1–4.

Hopley, D. 1988, 'Anthropogenic Influences on the Great Barrier Reef', *Australian Geographer*, vol. 19, no. 1, pp. 26–45.

Hopley, D., Parnell, K. E. and Isdale, P. J. 1989, 'The Great Barrier Reef Marine Park: Dimensions and Regional Patterns', *Australian Geographical Studies*, vol. 27, no. 1, pp. 47–66.

Horstman, M. 1992, 'Cape York Peninsula: Forging a Black–Green Alliance, *Habitat*, vol. 20, no. 2, pp. 18–25.

Horwitz, P. and Calver, M. 1998, 'Credible Science? Evaluating the Regional Forest Agreement Process in Western Australia', *Australian Journal of Environmental Management*, vol. 5, no. 4, pp. 213–25.

Howitt, R. 1989, 'Resource Development and Aborigines: The Case of Roeburne, 1960–1980', *Australian Geographical Studies*, vol. 27, no. 2, pp. 155–69.

Humphreys, J. S. and Walmsley, D. J. 1991, 'Locational Conflict in Metropolitan Areas: Melbourne and Sydney 1989', *Australian Geographical Studies*, vol. 29, no. 2, pp. 313–28.

Hundloe, T. and Neumann, R. (eds) 1988, *Environmental Practice in Australia*, Environment Institute of Australia/Griffith University, Brisbane.

Hutton, D. and Connors, L. 1999, *A History of the Australian Environmental Movement*, Cambridge University Press, Melbourne.

Institute of Australian Geographers 1999, 'Commentaries', *Australian Geographical Studies*, vol. 37, no. 3, pp. 300–42.

Isbell, R. F. 1986, 'The Tropical and Subtropical North and Northeast', in Russell, J. S. and Isbell, R. F. (eds), *Australian Soils: The Human Impact*, University of Queensland Press/Australian Soil Science Society, St Lucia, Qld, ch. 1, pp. 3–35.

Ivanovici, A., Tarte, D. and Olson, M. (eds) 1991, *Protection of Marine and Estuarine Areas: A Challenge for Australians*, Proceedings of the 4th Fenner Conference on Environment, Australian Committee for IUCN/Australian National Parks and Wildlife Service, Canberra.

Jakeman, A. J. and Simpson, R. W. 1987, *Air Quality and Resource Development*, Centre for Resource and Environmental Studies, Australian National University, Canberra.

Jeans, D. N. 1972, *An Historical Geography of New South Wales to 1901*, Reed Education, Sydney.
—— 1977, *Australia: A Geography*, Sydney University Press, Sydney.
—— 1983, 'Wilderness, Nature and Society: Contributions to the History of an Environmental Attitude', *Australian Geographical Studies*, vol. 21, no. 2, pp. 170–82.
—— (ed.) 1986, *Australia: A Geography*, vol. 1, *The Natural Environment*, Sydney University Press, Sydney.
—— (ed.) 1987, *Australia: A Geography*, vol. 2, *Space and Society*, Sydney University Press, Sydney
Jenkin, J. J. 1986, 'Western Civilisation', in Russell, J. S. and Isbell, R. F. (eds), *Australian Soils: The Human Impact*, University of Queensland Press/ Australian Soil Science Society, St Lucia, Qld, ch. 6, pp. 134–56.
Jenkins, K. 1997, 'Free Flow', *Ecos*, vol. 93, spring, pp. 10–13.
Jensen, H. I. 1914, *The Soils of New South Wales*, NSW Government Printer, Sydney.
Johnson, A. K. L., McDonald, G. T., Shrubsole, D. A. and Walker, D. H. 1998, 'Natural Resource Use and Management in the Australian Sugar Industry', *Australian Journal of Environmental Management*, vol. 5, no. 2, pp. 97–108.
Johnston, S. W. and Good, R. B. 1996, 'The Impact of Exogenous Zinc on the Soils and Plant Communities of Carruthers Peak, Kosciusko National Park NSW', *Proceedings ASSSI and NZSSS National Soils Conference*, Melbourne, vol. 3, pp. 117–18.
Jolly, P. B. and Chin, D. N. 1992, 'Hydrogeological Modelling for an Arid-zone Borefield in Amadeus Basin Aquifers, Alice Springs, Northern Territory', *BMR Journal of Australian Geology and Geophysics*, vol. 13, pp. 61–6.
Jones, B. P. 1992, 'UN Convention on Climate Change: Effects on Australia's Energy Sector', *Agriculture and Resources Quarterly*, vol. 4, no. 2, pp. 186–95.
Jones, G. O. 1990, 'Salt Movement Through Kerang Region Soils', pp. 61–70, in Humphreys, E., Muirhead, W. A. and van der Lelij, A. (eds), *Management of Soil Salinity in South East Australia*, Proceedings of a symposium held 18–20 September 1989, Albury, Australian Society of Soil Science, Riverina Branch, Wagga Wagga.
Jones, M. M. 1991, *Marine Organisms Transported in Ballast Water: A Review of Australia's Scientific Position*, Bureau of Rural Resources Bulletin No. 11, AGPS, Canberra.
Kailola, P., Williams, M., Ward, P. and Reichelt, R. 1993, 'Making the Most of Australia's Diverse Fisheries Resource', *Resource Sciences Interface*, September, pp. 14–21.
Kapp, W. A. 1980, 'A Study of Mine Subsidence at Two Collieries in the Southern Coalfield, New South Wales', *Proceedings, Australasian Institute of Mining and Metallurgy*, vol. 276, pp. 1–11.
—— 1982, 'Subsidence From Deep Longwall Mining of Coal Overlain by Massive Sandstone Strata', *Proceedings, Australasian Institute of Mining and Metallurgy*, vol. 281, pp. 23–36.

Kay, D. and Carter, J. P. 1992, 'Effects of Subsidence on Steep Topography and Cliff Lines, in Aziz, N. I. and Peng, S. S. (eds), *11th International Conference on Ground Control in Mining*, University of Wollongong, 7–10 July 1992, Australasian Institute of Mining and Metallurgy, Melbourne, pp. 483–90.

Kelly, R. 1992, *Report on the Earth Summit*, AGPS, Canberra.

Kenchington, R. A. 1990, *Managing Marine Environments*, Taylor & Francis, New York.

Kenworthy, J. 1987, 'Australian Cities: Beyond the Suburban Dream', *Habitat*, vol. 15, no. 6, pp. 3–7.

Kirkpatrick, J. 1988, 'Heritage and Development in Tasmania', *Australian Geographer*, vol. 19, no. 1, pp. 46–63.

—— 1994, *A Continent Transformed: Human Impact on the Natural Vegetation of Australia*, Oxford University Press, Melbourne.

—— 1998, 'Nature Conservation and the Regional Forest Agreement Process', *Australian Journal of Environmental Management*, vol. 5, no. 1, pp. 31–7.

Kirkup, H., Brierley, G., Brooks, A. and Pitman, A. 1998, 'Temporal Variability of Climate in South-eastern Australia: A Reassessment of Flood- and Drought-dominated Regimes', *Australian Geographer*, vol. 29, no. 2, pp. 241–55.

Knighton, A. D. 1987, 'Tin Mining and Sediment Supply to the Ringarooma River, Tasmania, 1875–1979', *Australian Geographical Studies*, vol. 25, no. 1, pp. 83–97.

Koch, J. M. and Ward, S. C. 1994, 'Establishment of Understorey Vegetation for Rehabilitation of Bauxite-Mined Areas in the Jarrah Forest of Western Australia', *Journal of Environmental Management*, no. 41, pp. 1–15.

Kynaston, E. 1981, *A Man on Edge: A Life of Baron Sir Ferdinand von Mueller*, Allen Lane/Penguin, Melbourne.

Lambert, I. B. and Perkin, D. J. 1998, 'Australia's Mineral Resources and Their Global Status', *AGSO Journal of Australian Geology and Geophysics*, vol. 17, no. 4, pp. 1–14.

Landsberg, J., Morse, J. and Khanna, P. 1990, 'Tree Dieback and Insect Dynamics in Remnants of Native Woodlands on Farms', in Saunders, D. A., Hopkins, A. J. M. and How, R. A. (eds), *Australian Ecosystems: 200 Years of Utilization, Degradation and Reconstruction*, Proceedings, Ecological Society of Australia, vol. 16, Surrey Beatty and Sons, Sydney, pp. 149–65.

Lawrence, D. and Dight, I. J. 1991, 'The Torres Strait Baseline Study: Environmental Protection of a Tropical Marine Environment in Northern Australia', in *Coastal Zone '91: 7th Symposium on Coastal & Ocean Management*, Long Beach, California, American Society of Civil Engineers, New York pp. 1125–39.

Lefroy, T. and Hobbs, R. 1992, 'Ecological Indicators For Sustainable Agriculture, *Australian Journal of Soil and Water Conservation*, vol. 5, no. 4, pp. 22–8.

Lenzen, M. 1998, 'Primary Energy and Greenhouse Gases Embodied in Australian Final Consumption: An Input–Output Analysis', *Energy Policy*, vol. 26, no. 6, pp. 495–506.

Lesslie, R. 1991, 'Wilderness Survey and Evaluation in Australia', *Australian Geographer*, vol. 22, no. 1, pp. 35–43.

Lindenmayer, D. 1994, 'Timber Harvesting Impacts on Wildlife: Implications For Ecologically Sustainable Forest Use, *Australian Journal of Environmental Management*, no. 1, pp. 56–68.

Lines, W. J. 1991, *Taming the Great South Land*, Allen & Unwin, Sydney.

Lloyd, D. 1992, 'Gunwarrie's Guns Solve the Salt Problems', *Australian Journal of Soil and Water Conservation*, vol. 5, no. 4, pp. 9–13.

Lockie, S. 1992, 'Landcare Before the Flood', *Rural Society*, vol. 2, no. 2, pp. 7–9.

Lovering, J. F. and Prescott, J. R. V. 1979, *Last of Lands: Antarctica*, Melbourne University Press, Melbourne.

Lunt, I. D. 1997a, 'The Distribution and Environmental Relationships of Native Grasslands on the Lowland Gippsland Plain, Victoria: An Historical Study', *Australian Geographical Studies*, vol. 35, no. 2, pp. 140–52.

—— 1997b, 'Tree Densities Last Century on the Lowland Gippsland Plain, Victoria', *Australian Geographical Studies*, vol. 35, no. 3, pp. 342–8.

Lyons, T., Kenworthy, J. and Newman, P. 1990, 'Urban Stucture and Air Pollution', *Atmospheric Environment*, vol. 24B, no. 1, pp. 43–8.

McDonald, P. 1992, 'Water Resources Development in the Northern Territory's Arid Zone', *BMR Journal of Australian Geology and Geophysics*, vol. 13, pp. 67–74.

McGarity, J. W. and Storrier, R. R. 1986, 'Fertilizers', in Russell, J. S. and Isbell, R. F. (eds), *Australian Soils: The Human Impact*, University of Queensland Press/Australian Soil Science Society, St Lucia, ch. 12, pp. 304–33.

McGarry, D. 1993, 'Degradation of Soil Structure', in McTainsh, G. H. and Boughton, W. C. (eds), *Land Degradation Processes in Australia*, Longman Cheshire, Melbourne, ch. 9, pp. 271–305.

McIlroy, D. (ed.) 1990, *The Ecological Future of Australia's Forests*, Australian Conservation Foundation, Melbourne.

Mackay, N. and Eastburn, D. (eds) 1990, *The Murray*, Murray Darling Basin Commission, Canberra.

MacKinnon, M. R. and Herbert, B. W. 1996, 'Temperature, Dissolved Oxygen and Stratification in a Tropical Reservoir, Lake Tinaroo, Northern Queensland, Australia', *Marine and Freshwater Research*, vol. 47, pp. 937–49.

McKinnon, K. 1993, *Review of Marine Research Organisations: Report to the Minister for Science and Small Business*, University of Wollongong, Wollongong, NSW.

McLoughlin, K. 1992, How Many Fish in the Sea?, *Rural Resources Interface*, no. 3, pp. 17–21.

McMahon, T. A., Gan, K. C. and Finlayson, B. L. 1992, 'Anthropogenic Influences to the Hydrologic Cycle in Australia', in Gifford, R. M. and

Barson, M. M. (eds), *Australia's Renewable Resources: Sustainability and Global Change*, Bureau of Rural Resources Proceedings, no. 14, pp. 36–66.

McMurtrie, R. E. and Dewar, R. C. 1997, 'Sustainable Forestry: A Model of the Effects of Nitrogen Removals in Wood Harvesting and Fire on the Nitrogen Balance of Regrowth Eucalypt Stands', *Australian Journal of Ecology*, vol. 22, pp. 243–55.

Macquarie Illustrated World Atlas, 1984, Macquarie Library, Sydney.

McTainsh, G. H. and Boughton, W. C. 1993, *Land Degradation Processes in Australia*, Longman Cheshire, Melbourne.

McTainsh, G. H. and Leys, J., 1993 'Soil Erosion by Wind', in McTainsh, G. H. and Boughton, W. C. (eds), *Land Degradation Processes in Australia*, Longman Cheshire, Melbourne, ch. 7, pp. 189–223.

McTainsh, G. H., Lynch, A. W. and Burgess, R. C. 1990, 'Wind Erosion in Eastern Australia', *Australian Journal of Soil Research*, no. 28, pp. 323–39.

Macumber, P. 1990, 'The Salinity Problem', in Mackay, N. and Eastburn, D. (eds), *The Murray*, Murray Darling Basin Commission, Canberra, ch. 7, pp. 111–25.

Main, B. Y. 1990, 'Restoration of Biological Scenarios: The Role of Museum Collections', in Saunders, D. A., Hopkins, A. J. M. and How, R. A. (eds), *Australian Ecosystems: 200 Years of Utilization, Degradation and Reconstruction*, Proceedings, Ecological Society of Australia, vol. 16, Surrey Beatty and Sons, Sydney, pp. 397–409.

Malcolm, C. V. 1983, *Wheatbelt Salinity*, Technical Bulletin no. 52, Western Australian Department of Agriculture, Perth.

Manins, P. J. 1986, *Air Quality in the Latrobe Valley*, Latrobe Valley Airshed Study Steering Committee, Environment Protection Authority, Melbourne.

Manner, H. I., Thaman, R. R. and Hassall, D. C. 1985, 'Plant Succession After Phosphate Mining on Nauru', *Australian Geographer*, vol. 16, no. 3, pp. 185–95.

Martin, H. A. 1987, 'Cainozoic History of the Vegetation and Climate of the Lachlan River Region, New South Wales', *Proceedings, Linnean Society of New South Wales*, vol. 109, no. 4, pp. 213–57.

Mercer, D. 1995, *A Question of Balance: Natural Resource Conflict Issues in Australia*, 2nd edn, The Federation Press, Sydney.

—— 1993, 'Victoria's *National Parks (Wilderness) Act 1992*: Background and Issues', *Australian Geographer*, vol. 24, no. 1, pp. 25–32.

Mercer, D., Keen, M. and Woodfull, J. 1994, 'Defining the Environmental Problem: Local Conservation Stratgeies in Metropolitan Victoria', *Australian Geographical Studies*, vol. 32, no. 1, pp. 41–57.

Mills, K. 1988, 'The Clearing of the Illawarra Rainforests: Problems in Reconstructing Pre-European Vegetation Patterns, *Australian Geographer*, vol. 19, no. 2, pp. 230–40.

Moffatt, I. 1992, 'The Evolution of the Sustainable Development Concept: A Perspective From Australia', *Australian Geographical Studies*, vol. 30, no. 1, pp. 27–42.

Moran, P. J. and Grant, T. R. 1993, 'Larval Settlement of Marine Fouling Organisms in Polluted Water from Port Kembla Harbour, Australia', *Marine Pollution Bulletin*, vol. 26, no. 9, pp. 512–14.

Moran, P., De'ath, G., Baker, V., Bass, D., Christie, C., Miller, I., Miller-Smith, B. and Thompson, A. 1992, 'Pattern of Outbreaks of Crown-of-Thorns Starfish (*Acanthaster planci L.*) Along the Great Barrier Reef Since 1966', *Australian Journal of Marine and Freshwater Research*, no. 43, pp. 555–68.

Morcom, L. A. and Westbrooke, M. E. 1998, 'The Pre-settlement vegetation of the Western and Central Wimmera Plains of Victoria, Australia', *Australian Geographical Studies*, vol. 36, no. 3, pp. 273–88.

Morgan, H. 1993, 'World Heritage Listings and the Threat to Sovereignty Over Land and its Use, *The Mining Review*, vol. 17, no. 5, pp. 26–8.

Morley, I. W. 1981, *Black Sands*, University of Queensland Press, St Lucia.

Morrison, R. 1988, *The Voyage of the Great Southern Ark*, Lansdowne Rigby, Sydney.

Morton, R. M. 1992, 'Fish Assemblages in Residential Canal Developments Near the Mouth of a Subtropical Queensland Estuary, *Australian Journal of Marine and Freshwater Research*, vol. 43, pp. 1359–71.

Mosley, J. G. 1972, 'Towards a History of Conservation', in Rapoport, A. (ed.), *Australia as a Human Setting*, Angus & Robertson, pp. 136–54.

—— 1989, 'History of Conservation of the Australian Alps', in Good, R. (ed), *The Scientific Significance of the Australian Alps, First Fenner Conference on the Environment, Canberra*, Australian Alps National Parks Liaison Committee/Australian Academy of Sciences, Canberra, ch. 20, pp. 345–56.

Mueck, S. G., Ough, K. and Banks, J. C. 1996, 'How Old Are Wet Forest Understories?', *Australian Journal of Ecology*, vol. 21, pp. 345–8.

Mulrennan, M. 1993, *Towards a Marine Strategy for Torres Strait (MaSTS)*, Australian National University Northern Australia Unit/Torres Strait Island Coordinating Council, Darwin.

Mulvey, P. 1992, 'Soil Cleanup Standards Do Not Exist in Australia', *Fulbright Symposium on Contaminated Sites in Australia: Challenges for Law and Public Policy*, University of New South Wales, Sydney.

Munro, C. H. 1974, *Australian Water Resources and their Development*, Angus & Robertson, Sydney.

Murray, S. 1976, 'Contemporary Architecture and Environmental Impact', in Seddon, G. and Davis, M. (eds), *Man and Landscape in Australia*, Australian Government Publishing Service, Canberra.

Nanson, G. C. and Erskine, W. D. 1988, 'Episodic Changes of Channels and Floodplains on Coastal Rivers in New South Wales', in Warner, R. F. (ed.), *Fluvial Geomorphology in Australia*, Academic Press, Sydney, pp. 201–21.

National Environment Protection Council 1999a, 'News Release: National Strategy to Combat Environmental Impacts of Motor Vehicles', released 3 March 1999, <http://www.nepc.gov.au/news.html> 15 May 1999.

—— 1999b, 'Assessment of Site Contamination: Draft National Environment Protection Measure and Impact Statement', released 12 March 1999, <http://www.nepc.gov.au/> 15 May 1999.

Neck, M., Barnes, P. and Cox, A. 1992, 'Greenhouse Policies and the Australian Aluminium Industry', *Agriculture and Resources Quarterly*, vol. 4, no. 3, pp. 345–54.

New South Wales Department of Mines 1901, *Report*, NSW Government Printer, Sydney.

New South Wales Pulp and Paper Industry 1989, *Task Force Report*, Department of State Development, Sydney.

Nichols, P. D., Leeming, R. and Cresswell, G. 1992, 'Sydney Deep Ocean Outfall Studies', *Chemistry in Australia*, vol. 59, no. 8, p. 389.

Noble, A. 1904, 'Dust in the Atmosphere During 1902–3', *Monthly Weather Review*, August, pp. 364–5.

Noble, J. C. and Tongway, D. J. 1986a, 'Pastoral Settlement in Arid and Semi-arid Rangelands', in Russell, J. S. and Isbell, R. F. (eds), *Australian Soils: The Human Impact*, University of Queensland Press/Australian Soil Science Society, St Lucia, ch. 9, pp. 217–42.

—— 1986b, 'Herbivores in Arid and Semi-arid Rangelands', in Russell, J. S. and Isbell, R. F. (eds), *Australian Soils: The Human Impact*, University of Queensland Press/Australian Soil Science Society, St Lucia, ch. 10, pp. 243–70.

North West Shelf Joint Ventures 1993, *North West Shelf Report*, no. 8, North West Shelf Joint Ventures, Perth.

Northcote, K. H. 1986, 'Soils', in Jeans, D. N. (ed.), *Australia: A Geography*, vol. 1, *The Natural Environment*, Sydney University Press, Sydney, ch. 9, pp. 223–59.

NSW Health Department 1996, *Proceedings of the Health and Urban Air Quality in NSW Conference, Sydney, 3–4 June 1996*, State Health Publication (EHB) 980036.

O'Brien, M. and Purdie, R. 1990, 'Identifying the National Estate: Issues and Directions in the Assessment of Significance of Places', in Saunders, D. A., Hopkins, A. J. M. and How, R. A. (eds), *Australian Ecosystems: 200 Years of Utilization, Degradation and Reconstruction*, Proceedings, Ecological Society of Australia, vol. 16, Surrey Beatty and Sons, Sydney, pp. 491–504.

O'Connor, B., Cameron, I. and Martin, D. 1990, 'Correlation Between Petrol Lead Additive Consumption and Atmospheric Lead Concentrations in Perth, Western Australia', *Atmospheric Environment*, vol. 24B, no. 3, pp. 413–17.

Page, K. J. and Carden, Y. R. 1998, 'Channel Adjustment Following the Crossing of a Threshold: Tarcutta Creek, Southeastern Australia', *Australian Geographical Studies*, vol. 36, no. 3, pp. 289–311.

Pease, M. I., Nethery, A. G. and Young, A. R. M. 1997, 'Acid Sulfate Soils and Acid Drainage, Lower Shoalhaven Floodplain, New South Wales', *Wetlands (Australia)*, vol. 16, no. 2, pp. 56–71.

Pick, J. H. 1942, *Australia's Dying Heart: Soil Erosion in the Inland*, Melbourne University Press, Melbourne.

Pigram, J. J. 1986, *Issues in the Management of Australia's Water Resources*, Longman Cheshire, Melbourne.

—— 1988, 'The Taming of the Waters', in Heathcote, R. L. (ed.), *The Australian Experience: Essays in Land Settlement and Resource Management*, Longman Cheshire, Melbourne, ch. 12, pp. 151–62.

Porritt, J. 1991, 'One World: The Science and Politics', *Australian Natural History*, vol. 24, no. 1, pp. 35–41.

Porter, C. F. 1985, *Environmental Impact Assessment: A Practical Guide*, University of Queensland Press/Australian Soil Science Society, St Lucia, Queensland.

Powell, J. M. 1976, *Environmental Management in Australia 1788–1914*, Oxford University Press, Melbourne.

—— 1977, *Mirrors of the New World*, Archon Books, Hamdon, Connecticut.

—— 1988, *An Historical Geography of Modern Australia*, Cambridge University Press, Melbourne.

Powerline Review Panel 1989, *Final Report*, Victorian Government Printer, Melbourne.

Pratley, J. and Robertson, A. (eds) 1998, *Agriculture and the Environmental Imperative*, CSIRO Publishing, Collingwood.

Preen, A. 1991, 'Oil and the Great Barrier Reef: Are We Prepared?', *Search*, vol. 22, no. 7, pp. 223–5.

Preston, C. A. 1995, 'The Impact of Urbanisation on Water Quality in Lane Cove Catchment: A Comparison of Urban and Non-urban Catchments', *Australian Geographical Studies*, vol. 33, pp. 19–30.

Pringle, A. W. (née Phillips) 1991, 'Fluvial Sediment Supply to the North Queensland Coast, Australia', *Australian Geographical Studies*, vol. 29, no. 1, pp. 114–38.

—— 1996, 'History, Geomorphological Problems and Effects of Dredging in Cleveland Bay, Queensland', *Australian Geographical Studies*, vol. 34, no. 1, pp. 58–80.

Prosser, I. 1991, 'A Comparison of Past and Present Episodes of Gully Erosion at Wangrah Creek, Southern Tablelands, New South Wales', *Australian Geographical Studies*, vol. 29, no. 1, pp. 139–54.

Pyke, G. H. and O'Connor, P. J. 1991, *Wildlife Conservation in the South-east Forests of New South Wales*, Technical Report of the Australian Museum No. 5, Australian Museum, Sydney.

Pyne, S. 1991, *Burning Bush: A Fire History of Australia*, Allen & Unwin, Sydney.

Ranger Uranium Environmental Inquiry 1976, *First Report*, AGPS, Canberra.

—— 1977, *Second Report*, AGPS, Canberra.

Recher, H. F., Shields, J., Kavanagh, R. and Webb, G. 1987, 'Retaining Remnant Mature Forest for Nature Conservation at Eden, NSW: A Review of Theory and Practice', in Saunders, D. A., Arnold, G. W., Burbidge, A. A. and Hopkins, A. J. M. (eds), *Nature Conservation: The Role of Remnants of Native Vegetation*, Surrey Beatty & Sons, Sydney.

Reid, A. F. 1997, 'Urban Air Pollution in Australia: Do We Have a Problem?', in Australian Medical Association, *Air Pollution and Health: The Facts*, Conference Papers, 10 November, Barton.

Reinfelds, I., Rutherfurd, I. and Bishop, P. 1995, 'History and Effects of Channelisation on the Latrobe River, Victoria', *Australian Geographical Studies*, vol. 33, no. 1, pp. 60–76.

Resource Assessment Commission 1992a, *Forest and Timber Inquiry Final Report*, Overview, AGPS, Canberra.

—— 1991a, *Kakadu Conservation Zone Inquiry, Final Report*, vol. 2, AGPS, Canberra.

—— 1991b, *Kakadu Conservation Zone Inquiry, Final Report*, vol. 1, AGPS, Canberra.

—— 1992b, *Forest and Timber Inquiry Final Report*, vol. 1, AGPS, Canberra.

—— 1992c, *Forest and Timber Inquiry Final Report*, vol. 2A, AGPS, Canberra.

—— 1992d, *A Survey of Australia's Forest Resources*, AGPS, Canberra.

—— 1993, *Resources and Uses of the Coastal Zone*, Coastal Zone Inquiry Information Paper No. 3, AGPS, Canberra.

Reynolds, R. G. 1977, *Coal Mining Under Stored Water*, Report of an Inquiry.

Rich, D. C. and Young, R. W. (eds) 1988, *Environment and Development in Australia*, Geographical Society of New South Wales, Sydney.

Richardson, H. H. 1978 [1917], *Australia Felix*, Penguin Books Australia, Melbourne.

Riley, S., Luscombe, G. and Williams, A. 1986, 'Urban Stormwater Design: Some Lessons From the 8 November 1984 Sydney Storm', *Australian Geographer*, vol. 17, no. 1, pp. 40–50.

Roberts, D. E. 1996, 'Patterns in Subtidal Assemblages Associated with a Deep-water Sewage Outfall', *Journal of Marine and Freshwater Research*, vol. 47, pp. 1–9.

Roberts, J. 1995, 'Evaporation Basins Are Wetlands', *Australian Journal of Environmental Management*, vol. 2, March, pp. 7–18.

Roberts, M. 1992, 'Costs and Trade-offs in Environmental Management', *Abstracts, Enviromine Australia Conference, Taronga Zoo, 13–22 March 1992*, Australasian Institute of Mining and Metallurgy/Zoological Parks Board of New South Wales, Sydney.

Rose, C. W. 1993, 'Soil Erosion by Water', in McTainsh, G. H. and Boughton, W. C. (eds), *Land Degradation Processes in Australia*, Longman Cheshire, Melbourne, ch. 6, pp. 149–87.

Rose, S. 1997, 'Influence of Suburban Edges on the Invasion of *Pittosporum undulatum* into the Bushland of Northern Sydney', *Australian Journal of Ecology*, vol. 22, pp. 89–99.

Rosewell, C. J. 1997, *Potential Sources of Sediments and Nutrients: Sheet and Rill Erosion and Phosphorus Sources*, Australia: State of the Environment Technical Paper Series (Inland waters), Department of Environment Sport and Territories, Canberra.

Routley, R. and Routley, V. 1974, *Fight for the Forests*, Research School of Social Sciences, Australian National University, Canberra.

Russell, J. S. and Isbell, R. F. 1986, *Australian Soils: The Human Impact*, University of Queensland Press, St Lucia.

Rutland, R. W. R. 1997, 'The Sustainability of Mineral Use', *AGSO Journal of Australian Geology and Geophysics*, vol. 17, no. 1, pp. 13–25.

Saueracker, G. 1991, 'The Abrolhos Islands: Keep Your Eyes Open', *Geo*, vol. 12, no. 4, pp. 18–25.

Saunders, D. A., Hopkins, A. J. M. and How, R. A. (eds) 1990, *Australian Ecosystems: 200 Years of Utilization, Degradation and Reconstruction*, Proceedings, Ecological Society of Australia, vol. 16, Surrey Beatty and Sons, Sydney.

Searle, B. 1997, The Environmental Impacts of Subsidence due to Underground Coal Mining in Darawal SRA, NSW, B.Env.Sc. (Hons) thesis, University of Wollongong.

Senate Standing Committee on Natural Resources 1985, *The Natural Resources of the Australian Antarctic Territory*, AGPS, Canberra.

Sim, R. 1993, The Contribution of Urban Runoff to Surface Water Quality in the Botany Wetlands, M.Sc. (Hons) thesis, University of Wollongong.

Simmons, P., Poulter, D. and Hall, N. 1991, '*Management of Irrigation Water in the Murray-Darling Basin*, Australian Bureau of Agricultural and Resource Economics Discussion Paper No. 91.6, Australian Bureau of Agricultural and Resource Economics, Canberra.

Smith, D. I. 1998, *Water in Australia: Resources and Management*, Oxford University Press, Melbourne.

Speirs, H. 1981, *Landscape Art and the Blue Mountains*, Griffin Press for Alternative Publishing Cooperative Ltd, Sydney.

Spencer, C., Robertson, A. I. and Curtis, A. 1998, 'Development and Testing of a Rapid Appraisal Wetland Condition Index in South-eastern Australia', *Journal of Environmental Management*, vol. 54, pp. 143–59.

State Forests of New South Wales 1994, *Environmental Impact Statement: Proposed Forestry Operations in Eden Management Area*, State Forests of New South Wales, Pennant Hills, NSW.

State Pollution Control Commission (NSW) 1980, *Pollution Control in the Hunter Valley with Particular Reference to Aluminium Smelting*, 2nd edn, SPCC, Sydney.

—— 1990, *Water Quality Criteria for New South Wales: Discussion Paper*, SPCC, Sydney.

—— 1991, *Coastal Resource Atlas for Oil Spills from Cape Dromedary to Cape Howe*, SPCC, Sydney.

Stromberg, J., Anderson, L., Bjork, G., Bonner, W., Clark, A., Dick, A., Ernst, W., Limbert, D., Peel, D., Priddle, J., Smith, R. and Walton, D. 1990, *State of the Marine Environment in Antarctica*, UNEP Regional Seas Reports and Studies No. 129, United Nations Environment Programme, Nairobi.

Sydney Water 1995, *Demand Management Strategy, October 1995*, Sydney Water Corporation, 64 pp.

Taylor, S. G. 1990, 'Naturalness: The Concept and its Application to Australian Ecosystems', in Saunders, D. A., Hopkins, A. J. M. and How, R. A. (eds), *Australian Ecosystems: 200 Years of Utilization, Degradation and Reconstruction*, Proceedings, Ecological Society of Australia, vol. 16, Surrey Beatty and Sons, Sydney, pp. 411–18.

Thackway, R. and Brunckhorst, D. J. 1998, 'Alternative Futures for Indigenous Cultural and Natural Areas in Australia's Rangelands', *Australian Journal of Environmental Management*, vol. 5, September, pp. 169–81.

Thomas, W. L. (ed.) 1956, *Man's Role in Changing the Face of the Earth*, University of Chicago Press, Chicago.

Thompson, H. 1990, *Mining and the Environment in Papua New Guinea*, Murdoch University Economics Programme Working Paper No. 39, Murdoch University, Perth.

Tiller, K. G. 1992, 'Urban Soil Contamination in Australia', *Australian Journal of Soil Research*, no. 30, pp. 937–57.

Twigg, L. E. and Fox, B. J. 1991, 'Recolonization of Regenerating Open Forest by Terrestrial Lizards Following Sand Mining', *Australian Journal of Ecology*, vol. 16, no. 2, pp. 137–48.

Underwood, A. 1991, 'Ecological Research and the Management of Human Exploitation on the Rocky Coast of New South Wales', in Ivanovici, A., Tarte, D. and Olson, M. (eds), *Protection of Marine and Estuarine Areas: A Challenge for Australians, Proceedings of 4th Fenner Conference on Environment*, Australian Committee for IUCN/Australian National Parks and Wildlife Service, Canberra, ch. 18, pp. 122–8.

van Oosterzee, P. 1984, 'The Recreation Opportunity Spectrum: Its Use and Misuse', *Australian Geographer*, vol. 16, no. 2, pp. 97–104.

Van Saane, L. and Gordon, B. 1991, 'Management and Planning of Tasmania's Native Forests for Sawlog and Pulpwood', in McKinnell, F. H., Hopkins, E. R. and Fox, J. E. D. (eds), *Forest Management in Australia*, Surrey Beatty and Sons, Sydney, ch. 4, pp. 58–76.

Vellacott, S. 1992, 'Progress in the Development of a Water-care Ethic for the Pilbara Region of Western Australia', *BMR Journal of Australian Geology and Geophysics*, vol. 13, pp. 75–82.

Wace, N. 1994, 'Beachcombing for Ocean Litter', *Australian Natural History*, vol. 24, no. 9, pp. 46–52.

Walker, T. A., Bell, P. R. F., Gabric, A. J., and Kinsey, D. W. 1991, 'Opinions: Pollution and the Great Barrier Reef', *Search*, vol. 22, no. 4, pp. 115–21.

Warner, R. F. 1983, 'Channel Changes in the Sandstone and Shale Reaches of the Nepean Valley, New South Wales', in Young, R. W. and Nanson, G. C. (eds), *Aspects of Australian Sandstone Landscapes*, Australian and New Zealand Geomorphology Group, Wollongong, pp. 106–19.

—— 1986, 'Hydrology', in Jeans, D. N. (ed.), *Australia: A Geography*, vol. 1, *The Natural Environment*, Sydney University Press, Sydney, ch. 2, pp. 49–79.

—— 1991, 'Impacts of Environmental Degradation on Rivers, With Some Examples from the Hawkesbury–Nepean System', *Australian Geographer*, vol. 22, no. 1, pp. 1–13.

Waterhouse, B. M. 1988, 'Broom (*Cytisus scoparius*) at Barrington Tops, New South Wales', *Australian Geographical Studies*, vol. 26, no. 2, pp. 239–48.

Waterhouse, J. D. and Armstrong, D. 1991, 'Operation and Management of the Olympic Dam Project Water Supply Scheme', in Australian Water Resources Council, *Proceedings, International Conference on Groundwater in Large Sedimentary Basins, Perth, 9–13 July 1990*, AGPS, Canberra, pp. 246–55.

Watson, C. L. 1986, 'Irrigation', in Russell, J. F. and Isbell, R. F. (eds), *Australin Soils: The Human Impact*, University of Queensland Press/Australian Soil Science Society, St Lucia, ch. 13, pp. 334–56.

Watson, I. 1990, *Fighting Over the Forests*, Allen & Unwin, Sydney.

Whinam, J., Cannell, E., Kirkpatrick, J. and Comfort, M. 1994, 'Studies on the Potential Impact of Recreational Horseriding on Some Alpine Environments of the Central Plateau, Tasmania', *Journal of Environmental Management*, no. 40, pp. 103–17.

Whitbread, A. M., Lefroy, R. D. B and Blair, G. J. 1998, 'A Survey of the Impact of Cropping on Soil Physical and Chemical Properties in North-western New South Wales', *Australian Journal of Soil Research*, vol. 36, pp. 669–81.

White, M. E. 1986, *The Greening of Gondwana*, Reed Books, Sydney.

—— 1997, *Listen Our Land Is Crying*, Kangaroo Press, Sydney.

Whitelock, D. 1985, *Conquest to Conservation*, Wakefield Press, Adelaide.

Whyte, R. J. and Conlon, M. L. 1990, *The New South Wales Cotton Industry and the Environment*, State Pollution Control Commission, Sydney.

Williams, J., Bui, E. N., Gardner, E. A., Littleboy, M. and Probert, M. E. 1997, 'Tree Clearing and Salinity Hazard in the Upper Burdekin Catchment of North Queensland', *Australian Journal of Soil Research*, vol. 35, pp. 785–801.

Williams, M. 1969, 'The Spread of Settlement in South Australia', in Gale, F. and Lawton, G. H. (eds), *Settlement and Encounter*, Oxford University Press, Melbourne, pp. 1–50.

—— 1974, *The Making of the South Australian Landscape*, Academic Press, London.

—— 1988, 'The Clearing of the Woods', in Heathcote, R. L. (ed.), *The Australian Experience: Essays in Land Settlement and Resource Management*, Longman Cheshire, Melbourne, ch. 9, pp. 115–26.

Williamson, D. R. 1983, 'The Application of Salt and Water Balances to Quantify Causes of the Dryland Salinity Problem in Victoria', *Proceedings, Royal Society of Victoria*, vol. 95, no. 3, pp. 103–11.

Wilson, B. A., Robertson, D., Moloney, D. J., Newell, G. R. and Ladlaw, W. S. 1990, 'Factors Affecting Small Mammal Distribution and Abundance in the Eastern Otway Ranges, Victoria', in Saunders, D. A., Hopkins, A. J. M. and How, R. A. (eds), *Australian Ecosystems: 200 Years of Utilization, Degradation and Reconstruction*, Proceedings, Ecological Society of Australia, vol. 16, Surrey Beatty and Sons, Sydney, pp. 379–96.

Wood, D. 1997, 'Limits Reaffirmed: New Wheat Frontiers in Australia 1916–1939', *Journal of Historical Geography*, vol. 24, no. 2, pp. 459–77.

Woodley, S., Craik, W. and Kelleher, G. 1990, 'Environmental Engineering in the Great Barrier Reef Marine Park', *Proceedings, Institution of Engineers (Australia) National Conference, Canberra*, The Institution of Engineers (Australia), Canberra, pp. 131–6.

Woods, L. E. 1984, *Land Degradation in Australia*, AGPS, Canberra.

Woolmington, E. 1972, 'The Australian Environment as a Problem Area', in Rapoport, A. (ed.), *Australia as a Human Setting*, Angus & Robertson, Sydney, pp. 22–36.

Working Party on Dryland Salting in Australia 1982, *Salting on Non-irrigated Land in Australia*, Soil Conservation Authority of Victoria for the Standing Committee on Soil Conservation, F. D. Atkinson Government Printer, Melbourne.

Wray, R. A. L., Young, R. W. and Price, D. M. 1993, 'Cainozoic Heritage in the Modern Landscape Near Bungonia, Southern New South Wales', *Australian Geographer*, vol. 24, no. 1, pp. 45–61.

Wright, A. 1991–92, 'Smog Moves West as Sydney Grows', *Ecos*, no. 70, summer, pp. 17–22.

Wright, J. 1977, *The Coral Battleground*, Nelson, Melbourne.

Yim, W. 1991, 'Tin Placer Genesis in Northeastern Tasmania', in Williams, M., De Deckker, P. and Kershaw, A. (eds), *The Cainozoic in Australia: A Re-appraisal of the Evidence*, Geological Society of Australia Special Publication No. 18, pp. 235–57.

Young, A., Marthick, J., Chafer, C. and Caldwell, J. 1996, 'Developing a Management Framework for Wetlands: The Illawarra Experience', *Australian Journal of Environmental Management*, vol. 3, no. 4, pp. 218–28.

Young, A. R. M., Bryant, E. A. and Winchester, H. P. M. 1992, 'The Wollongong Lead Study', *Australian Geographer*, vol. 23, no. 2, pp. 121–33.

Young, R. and McDougall, I. 1993, 'Long-term Landscape Evolution: Early Miocene and Modern Rivers in Southern New South Wales, Australia', *Journal of Geology*, vol. 101, pp. 35–49.

Young, R. W. 1987, 'Erodibility of Sandstone Surfaces of the Bungle Bungle Range, East Kimberley Region', Report to the Heritage Committee of Western Australia.

Young, R. W., Nanson, G. C. and Bryant, E. A. 1986, 'Alluvial Chronology for Coastal New South Wales: Climatic Control or Random Erosional Events', *Search*, vol. 17, nos 10–12, pp. 270–2.

Young, R. W. and Reffell, G. 1981, *Illawarra Region Sand Mining Study*, Department of Environment and Planning, Wollongong, NSW.

Young, R. W. and Young, A. R. M. 1988, ' "Altogether Barren, Peculiarly Romantic": The Sandstone Lands around Sydney', in Rich, D. C. and Young, R. W. (eds), *Environment and Development in Australia*, Geographical Society of New South Wales, Sydney, pp. 9–25.

—— 1992, *Sandstone Landforms*, Springer-Verlag, Berlin.

Yulsman, T. 1994, 'Lost in the Outback', *Earth*, vol. 3, no. 7, pp. 24–34.

Zann, P. (ed.) 1995, *Our Sea, Our Future*, Great Barrier Reef Marine Park Authority for Department of Arts Sport and Territories, Canberra.

Index

Note: Page references followed by an * indicate figures; those followed by a *t* indicate tables.